Selected Titles in This Series

723 **Deborah M. King and John B. Strantzen,** Maximum entropy of cycles of even period, 2001

722 **Hernán Cendra, Jerrold E. Marsden, and Tudor S. Ratiu,** Lagrangian reduction by stages, 2001

721 **Ingrid C. Bauer,** Surfaces with $K^2 = 7$ and $p_g = 4$, 2001

720 **Palle E. T. Jorgensen,** Ruelle operators: Functions which are harmonic with respect to a transfer operator, 2001

719 **Steve Hofmann and John L. Lewis,** The Dirichlet problem for parabolic operators with singular drift terms, 2001

718 **Bernhard Lani-Wayda,** Wandering solutions of delay equations with sine-like feedback, 2001

717 **Ron Brown,** Frobenius groups and classical maximal orders, 2001

716 **John H. Palmieri,** Stable homotopy over the Steenrod algebra, 2001

715 **W. N. Everitt and L. Markus,** Multi-interval linear ordinary boundary value problems and complex symplectic algebra, 2001

714 **Earl Berkson, Jean Bourgain, and Aleksander Pełczynski,** Canonical Sobolev projections of weak type $(1,1)$, 2001

713 **Dorina Mitrea, Marius Mitrea, and Michael Taylor,** Layer potentials, the Hodge Laplacian, and global boundary problems in nonsmooth Riemannian manifolds, 2001

712 **Raúl E. Curto and Woo Young Lee,** Joint hyponormality of Toeplitz pairs, 2001

711 **V. G. Kac, C. Martinez, and E. Zelmanov,** Graded simple Jordan superalgebras of growth one, 2001

710 **Brian Marcus and Selim Tuncel,** Resolving Markov chains onto Bernoulli shifts via positive polynomials, 2001

709 **B. V. Rajarama Bhat,** Cocylces of CCR flows, 2001

708 **William M. Kantor and Ákos Seress,** Black box classical groups, 2001

707 **Henning Krause,** The spectrum of a module category, 2001

706 **Jonathan Brundan, Richard Dipper, and Alexander Kleshchev,** Quantum Linear groups and representations of $GL_n(\mathbb{F}_q)$, 2001

705 **I. Moerdijk and J. J. C. Vermeulen,** Proper maps of toposes, 2000

704 **Jeff Hooper, Victor Snaith, and Min van Tran,** The second Chinburg conjecture for quaternion fields, 2000

703 **Erik Guentner, Nigel Higson, and Jody Trout,** Equivariant E-theory for C^*-algebras, 2000

702 **Ilijas Farah,** Analytic guotients: Theory of liftings for quotients over analytic ideals on the integers, 2000

701 **Paul Selick and Jie Wu,** On natural coalgebra decompositions of tensor algebras and loop suspensions, 2000

700 **Vicente Cortés,** A new construction of homogeneous quaternionic manifolds and related geometric structures, 2000

699 **Alexander Fel′shtyn,** Dynamical zeta functions, Nielsen theory and Reidemeister torsion, 2000

698 **Andrew R. Kustin,** Complexes associated to two vectors and a rectangular matrix, 2000

697 **Deguang Han and David R. Larson,** Frames, bases and group representations, 2000

696 **Donald J. Estep, Mats G. Larson, and Roy D. Williams,** Estimating the error of numerical solutions of systems of reaction-diffusion equations, 2000

695 **Vitaly Bergelson and Randall McCutcheon,** An ergodic IP polynomial Szemerédi theorem, 2000

(Continued in the back of this publication)

Maximum Entropy of Cycles of Even Period

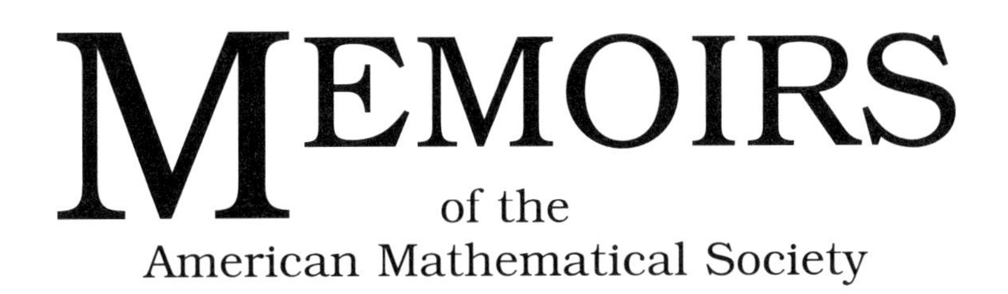

Number 723

Maximum Entropy of Cycles of Even Period

Deborah M. King
John B. Strantzen

July 2001 • Volume 152 • Number 723 (fourth of 5 numbers) • ISSN 0065-9266

American Mathematical Society
Providence, Rhode Island

2000 *Mathematics Subject Classification.*
Primary 37B40, 37E15.

Library of Congress Cataloging-in-Publication Data

King, Deborah M. (Deborah Martina), 1960–
Maximum entropy of cycles of even period / Deborah M. King, John B. Strantzen
p. cm. — (Memoirs of the American Mathematical Society, ISSN 0065-9266 ; no. 723)
"July 2001, Volume 152, Number 723 (fourth of 5 numbers)."
Includes bibliographical references
ISBN 0-8218-2707-3
1. Topological entropy. 2. Combinatorial dynamics. I. Strantzen, J. B. (John Bruce), 1942–
II. Title. III. Series.
QA3.A57 no. 723
[QA611.5]
510 s—dc21
[514] 2001023976

Memoirs of the American Mathematical Society

This journal is devoted entirely to research in pure and applied mathematics.

Subscription information. The 2001 subscription begins with volume 149 and consists of six mailings, each containing one or more numbers. Subscription prices for 2001 are $494 list, $395 institutional member. A late charge of 10% of the subscription price will be imposed on orders received from nonmembers after January 1 of the subscription year. Subscribers outside the United States and India must pay a postage surcharge of $31; subscribers in India must pay a postage surcharge of $43. Expedited delivery to destinations in North America $35; elsewhere $130. Each number may be ordered separately; *please specify number* when ordering an individual number. For prices and titles of recently released numbers, see the New Publications sections of the *Notices of the American Mathematical Society.*

Back number information. For back issues see the *AMS Catalog of Publications.*

Subscriptions and orders should be addressed to the American Mathematical Society, P. O. Box 845904, Boston, MA 02284-5904. *All orders must be accompanied by payment.* Other correspondence should be addressed to Box 6248, Providence, RI 02940-6248.

Memoirs of the American Mathematical Society is published bimonthly (each volume consisting usually of more than one number) by the American Mathematical Society at 201 Charles Street, Providence, RI 02904-2294. Periodicals postage paid at Providence, RI. Postmaster: Send address changes to Memoirs, American Mathematical Society, P. O. Box 6248, Providence, RI 02940-6248.

This publication is indexed in *Science Citation Index®*, *SciSearch®*, *Research Alert®*, *CompuMath Citation Index®*, *Current Contents®/Physical, Chemical & Earth Sciences.*
Printed in the United States of America.

♾ The paper used in this book is acid-free and falls within the guidelines established to ensure permanence and durability.
Visit the AMS home page at URL: `http://www.ams.org/`

10 9 8 7 6 5 4 3 2 1 06 05 04 03 02 01

Contents

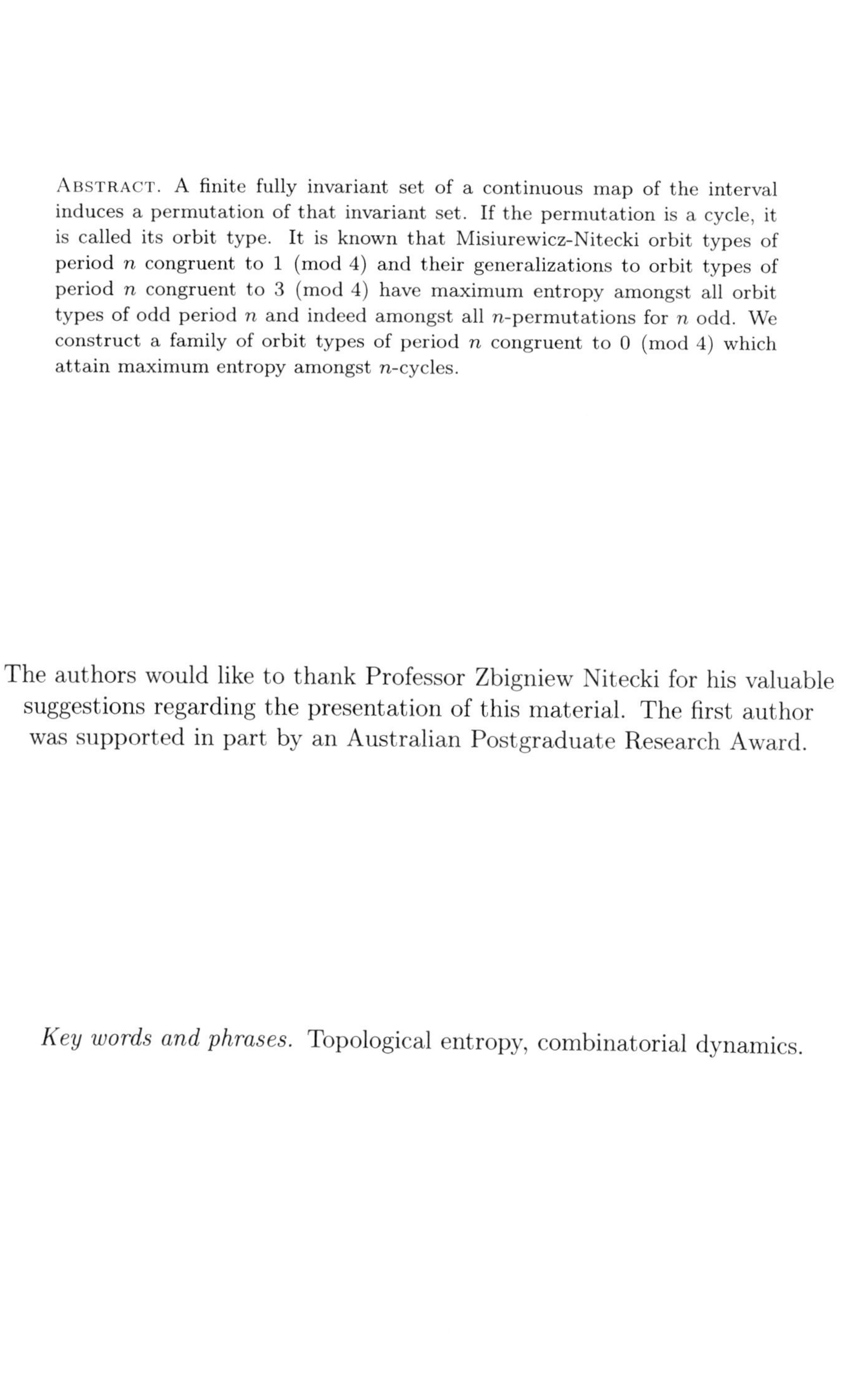

ABSTRACT. A finite fully invariant set of a continuous map of the interval induces a permutation of that invariant set. If the permutation is a cycle, it is called its orbit type. It is known that Misiurewicz-Nitecki orbit types of period n congruent to 1 (mod 4) and their generalizations to orbit types of period n congruent to 3 (mod 4) have maximum entropy amongst all orbit types of odd period n and indeed amongst all n-permutations for n odd. We construct a family of orbit types of period n congruent to 0 (mod 4) which attain maximum entropy amongst n-cycles.

The authors would like to thank Professor Zbigniew Nitecki for his valuable suggestions regarding the presentation of this material. The first author was supported in part by an Australian Postgraduate Research Award.

Key words and phrases. Topological entropy, combinatorial dynamics.

1. Introduction

In recent years much attention has been given to the question of finding n-permutations and n-cycles which attain maximum entropy amongst all n-permutations and n-cycles, respectively. Originally this question was posed by Misiurewicz and Nitecki [**10**] who showed that the maximum entropy for n-permutations and for n-cycles is asymptotic to $\log(2n/\pi)$. To obtain this result they defined a family of n-cycles for $n \equiv 1 \pmod 4$. Later, Geller and Tolosa [**4**] extended this definition to all n odd and proved that this family actually attains maximum entropy amongst all n-permutations (and hence all n-cycles). Geller and Weiss [**5**] then went on to show that this family is unique up to duality. For the case where n is even, King [**8, 9**] and Geller and Zhang [**6**] have shown, independently, that there are exactly two families of n-permutations (neither of which is a family of n-cycles), up to duality, which attain maximum entropy amongst n-permutations.

In this paper we construct a family of n-cycles which attain maximum entropy amongst all n-cycles of period $n \equiv 0 \pmod 4$.

2. Preliminaries

In this section we introduce some standard notation and well known results.

Let f be a continuous map of a compact interval I into itself. The *orbit* under f of a point $x \in I$ is the sequence $Orb(x) = \{x, f(x), f^2(x), \dots\}$ (note that $Orb(x)$ is an invariant set for f). If $x = f^m(x)$ for some $m \in \mathbb{N}$, then $Orb(x)$ is finite and the orbit is *periodic*.

Recall that a *permutation* on n letters is a bijective map $\theta : \{1, \dots, n\} \to \{1, \dots, n\}$. If θ has the property that for $1 \leq p \leq n$, $\theta^p(1) = 1$ if and only if $p = n$, then θ is a *cycle*.

We define P_n to be the set of all permutations on n letters and C_n to be the set of all cycles on n letters. We also let $P = \cup_{n \geq 1} P_n$ and $C = \cup_{n \geq 1} C_n$.

DEFINITION 2.1. Let $\theta \in P_n$. Then

1. the *dual* of θ is the permutation $\overline{\theta} \in P_n$ where $\overline{\theta}(i) = n + 1 - \theta(n + 1 - i)$, for $i \in \{1, \dots, n\}$,
2. the *reverse* of θ is the permutation $\widetilde{\theta} \in P_n$ where $\widetilde{\theta}(i) = \theta(n + 1 - i)$, for $i \in \{1, \dots, n\}$,
3. the dual of the reverse of θ is the permutation $\theta^* \in P_n$ where $\theta^*(i) = n + 1 - \theta(i)$, for $i \in \{1, \dots, n\}$,
4. the *flip* permutation $\varphi \in P_n$ is defined as $\varphi(i) = n + 1 - i$, for $i \in \{1, \dots, n\}$.

We can formulate $\widetilde{\theta}$, θ^* and $\overline{\theta}$ in terms of the flip permutation φ as follows:

$$
\begin{aligned}
\widetilde{\theta}(i) &= \theta(\varphi(i)), \\
\theta^*(i) &= \varphi(\theta(i)), \\
\overline{\theta}(i) &= \varphi(\theta(\varphi(i))) = \varphi(\widetilde{\theta}(i)) = \theta^*(\varphi(i)).
\end{aligned}
$$

If $\theta \in C_n$ then $\overline{\theta} \in C_n$ but $\widetilde{\theta}$ is not necessarily in C_n.

If x has a periodic orbit under f of period n we can write $Orb(x) = \{p_1, \dots, p_n\}$ with $p_1 < p_2 < \cdots < p_n$. This induces a cycle $\theta \in C_n$ in the following way:

$$\theta(i) = j \text{ if and only if } f(p_i) = p_j.$$

Received by the editor February 24, 1998, and in revised form April 21, 2000.

The cycle θ is called the *orbit type* of $Orb(x)$. More generally, if $S = \{p_1, \ldots, p_n\}$ (with $p_1 < p_2 < \cdots < p_n$) is any finite fully invariant set for f (that is, $f(S) = S$), we can define the *type* of S to be the permutation $\theta \in P_n$ with

$$\theta(i) = j \text{ if and only if } f(p_i) = p_j.$$

If S is a fully invariant set for f of type $\theta \in P_n$ then there is a unique map $f_\theta : [1, n] \longrightarrow [1, n]$ satisfying

(i) $f_\theta(i) = \theta(i)$, for $i \in \{1, \ldots, n\}$,

(ii) f_θ is affine on each interval $I_i = \{x \in \mathbb{R} : i \leq x \leq i + 1\}$, for $i \in \{1, \ldots, n - 1\}$.

Note that f_θ is also a map with an invariant set of type θ.

The map f_θ is called the *linearisation* of f with respect to its invariant set of type θ. Note that f_θ is entirely determined by θ. It is a particularly simple map amongst those with an invariant set of type θ. This also shows that every $\theta \in P_n$ is the type of an invariant set for some function f.

If, for each $i \in \{1, \ldots, n\}$, $f_\theta(i)$ is a local extremum of f_θ then f_θ is said to be *maximodal*. The permutation $\theta \in P_n$ is also said to be maximodal.

DEFINITION 2.2. The *entropy of a permutation* $\theta \in P$ is

$$h(\theta) = \inf \{h(f)\}$$

where f is a continuous map of a compact interval into itself which has an invariant set of type θ and $h(f)$ is the topological entropy of f (see [**2, pg 191**]).

The determination of the entropy of θ is markedly simplified by the following result which can be found in Misiurewicz and Szlenk [**11**]:

PROPOSITION 2.3. *If $\theta \in P$ then $h(\theta) = h(f_\theta)$.*

We note that for each $\theta \in P$, $h(\overline{\theta}) = h(\theta)$.

NOTATION 2.4. For $a, b \in \mathbb{R}$ with $1 \leq a \leq b \leq n - 1$,

$$\begin{aligned} [a, b] &= \{m \in \mathbb{N} : a \leq m \leq b\}; \\ O[a, b] &= \{m \in \mathbb{N} : m \textit{ is odd and } a \leq m \leq b\}; \\ E[a, b] &= \{m \in \mathbb{N} : m \textit{ is even and } a \leq m \leq b\}. \end{aligned}$$

NOTATION 2.5. For any $(n - 1) \times (n - 1)$ matrix M with non-negative entries $m_{i\,j}$ we let

$$\left|M^{(j)}\right| = \sum_{i=1}^{n-1} m_{i\,j}$$

and

$$\|M\| = \sum_{j=1}^{n-1} \sum_{i=1}^{n-1} m_{i\,j} = \sum_{j=1}^{n-1} \left|M^{(j)}\right|.$$

Further simplification regarding the determination of $h(\theta)$ is obtained as follows:

DEFINITION 2.6. The *induced matrix* $M(\theta)$ of $\theta \in P_n$ is the $(n-1) \times (n-1)$ matrix with $i\,j^{th}$ entry given by

$$a_{i\,j} = \begin{cases} 1, & \text{if } f_\theta(I_i) \supset I_j \\ 0, & \text{otherwise,} \end{cases}$$

where $I_i = \{x \in \mathbb{R} : i \leq x \leq i+1\}$ and $i, j \in [1, n-1]$.

PROPOSITION 2.7. *If $\theta \in P$ then $h(\theta) = \log \rho(M(\theta))$, where $\rho(M(\theta))$ is the spectral radius of the induced matrix of θ.*

The proof of Proposition 2.7 can be found in Block and Coppel [**2, Chapt VIII, Propn 19**].

We wish to identify a cycle $\theta_n \in C_n$ which has maximum entropy. The following discussion helps to eliminate some members of C_n which do not have maximum entropy.

If $\theta, \phi \in P$ we say that θ *forces* ϕ and we write $\theta \vdash \phi$ if every map which has an invariant set of type θ also has an invariant set of type ϕ.

PROPOSITION 2.8. *The relation $\vdash$ is a partial preorder on P and a partial order on C.*

That is, $\vdash$ is reflexive and transitive on P and in addition is anti-symmetric on C. The anti-symmetry property is due to Baldwin [**1**].

DEFINITION 2.9. The permutation θ is *forcing-maximal* on a subset S of P if $\theta \in S$ and, for all $\phi \in S$, if $\phi \vdash \theta$ then $\theta \vdash \phi$.

If $\theta, \phi \in P$ and $\theta \vdash \phi$ then, from Definition 2.2, clearly $h(\theta) \geq h(\phi)$.

Since entropy respects the partial order in C_n, we need only consider those cycles which are forcing-maximal. The next result, due to Jungreis [**7, Corollary 9.6**], allows us to analyse this class more simply.

THEOREM 2.10. *If the cycle $\theta \in C_n$ is forcing-maximal then it is maximodal.*

3. Some useful properties of the induced matrix of a maximodal permutation

The definition of the induced matrix $M(\theta)$ of a maximodal permutation θ naturally gives rise to easily assimilated information regarding the rows of the matrix. In particular, the following result is easy to show:

REMARK 3.1. Let $a_{i\,j}$ be the entries in the induced matrix of a maximodal permutation θ, where $i, j \in [1, n-1]$. If f_θ assumes a minimum at 1; that is, $\theta(1) < \theta(2)$, then

$$a_{i\,j} = 1 \iff \begin{cases} \theta(i) \leq j \leq \theta(i+1) - 1, & \text{if } i \text{ is odd} \\ \theta(i+1) \leq j \leq \theta(i) - 1, & \text{if } i \text{ is even.} \end{cases}$$

If f_θ has a maximum at 1; that is, $\theta(1) > \theta(2)$, then

$$a_{i\,j} = 1 \iff \begin{cases} \theta(i+1) \leq j \leq \theta(i) - 1, & \text{if } i \text{ is odd} \\ \theta(i) \leq j \leq \theta(i+1) - 1, & \text{if } i \text{ is even.} \end{cases}$$

We note that a trivial corollary of the above is that if $1 \leq i \leq n-1$, $1 \leq j_1 < j_2 \leq n-1$ and $a_{i\,j_1} = a_{i\,j_2} = 1$, then $a_{i\,j} = 1$ for $j_1 \leq j \leq j_2$; that is, there are no "gaps" in rows of 1's. This idea in conjunction with Remark 3.1 leads to the even more useful corollary:

LEMMA 3.2 (The Shape Lemma). *Let θ be a maximodal permutation and denote the entries of the induced matrix $M(\theta)$ by $a_{i\,j}$. Suppose that for some indices i, j we have*

$$a_{i-1\,j} = a_{i\,j} = 1.$$

1. *If $\theta(1) < \theta(2)$ then*

$$a_{i-1\,j'} = a_{i\,j'}$$

for every $j' \leq j$ (respectively, $j' \geq j$) if i is odd (respectively, i is even).

2. *If $\theta(1) > \theta(2)$ then*

$$a_{i-1\,j'} = a_{i\,j'}$$

for every $j' \leq j$ (respectively, $j' \geq j$) if i is even (respectively, i is odd).

PROOF. This is a direct consequence of Remark 3.1. To see how it works in the first of the four cases, assume that $\theta(1) < \theta(2)$, i is odd and $a_{i-1\,j} = a_{i\,j} = 1$ for some j. We wish to show that $a_{i-1\,j'} = a_{i\,j'}$ for every $j' \leq j$.

By Remark 3.1, since $\theta(1) < \theta(2)$ and i is odd we have

$$a_{i\,j'} = 1 \iff \theta(i) \leq j' \leq \theta(i+1) - 1$$

and since $i-1$ is even,

$$a_{i-1\,j'} = 1 \iff \theta(i-1+1) = \theta(i) \leq j' \leq \theta(i-1) - 1.$$

Furthermore, since $a_{i-1\,j} = a_{i\,j} = 1$ we have

$$\theta(i) \leq j \leq \theta(i+1) - 1 \quad \text{and} \quad \theta(i) \leq j \leq \theta(i-1) - 1.$$

Thus for each $j' \leq j$, either

$$\theta(i) \leq j' \ (\leq j) \text{ and so } a_{i-1\,j'} = a_{i\,j'} = 1$$

or

$$1 \leq j' < \theta(i) \text{ and so } a_{i-1\,j'} = a_{i\,j'} = 0.$$

The other three cases follow similarly. □

Remark 3.1 also permits us to retrieve θ from $M(\theta)$ by the following algorithm: If $\theta(1) < \theta(2)$ and i is odd, then $j = \theta(i)$ is found by scanning the row i from the left and stopping at the first 1, which corresponds to column j, whilst if i is even then $j = \theta(i)$ is found by scanning row i from the right and stopping one place before the first 1 (we note that this place will be n if $a_{i\,n-1} = 1$), which corresponds to column j. This also tells us that given columns j and $j-1$ of $M(\theta)$ for some $j \in [2, n-1]$ we can find $i = \theta^{-1}(j)$. (Given column 1 we can find $\theta^{-1}(1)$ and given column $n-1$ we can find $\theta^{-1}(n)$). An analogous algorithm holds if $\theta(1) > \theta(2)$.

Practically we also need information about the columns of $M(\theta)$. The Shape Lemma helps as it tells us about consecutive rows of $M(\theta)$. In fact a combination of the Shape Lemma and the fact that $\theta^{-1}(i)$ is unique, leads both to the well known result

COROLLARY 3.3. *Let n be even and $\theta \in P_n$ be maximodal with induced matrix $B = M(\theta)$. The jth column sum $\left|B^{(j)}\right|$ of B satisfies*

$$\left|B^{(j)}\right| \leq \begin{cases} 2j, & \text{if } j < 2k \\ n-1, & \text{if } j = 2k \\ 2(n-j), & \text{if } j > 2k \end{cases}$$

and the even stronger result

COROLLARY 3.4. *Let n be even and $\theta \in P_n$ be maximodal with induced matrix $B = M(\theta)$. If*

$$\left|B^{(j)}\right| = 2j \text{ for some } j < 2k \text{ then } \left|B^{(j')}\right| = 2j' \text{ for all } j' \leq j$$

and if

$$\left|B^{(j)}\right| = 2(n-j) \text{ for some } j > 2k \text{ then } \left|B^{(j')}\right| = 2(n-j) \text{ for all } j' \geq j.$$

Corollary 3.3 may also be obtained from the following result due to Misiurewicz and Nitecki [**10, Lemma 11.9**]:

PROPOSITION 3.5. *If θ is a permutation of length n, then its induced matrix $M = M(\theta)$, of order $(n-1) \times (n-1)$, has j^{th} column sum*

$$\left|M^{(j)}\right| \leq \min\{2j,\ 2(n-j)\}.$$

4. The family of orbit types

In this section we define the orbit type θ_n of period $n = 4k$, $k > 1$, and state some general features of θ_n and its induced matrix $M(\theta_n)$. We will show that this family of orbit types achieves maximum entropy amongst all orbit types of period n. In the case $n = 4$, the cycle which achieves maximum entropy also achieves maximum entropy amongst 4-permutations and so is not considered here (see King [**8**]).

DEFINITION 4.1. We define θ_n by

$$\theta_n : \quad j \mapsto \begin{cases} 2k - j + 1, & \text{if } j \in O[1, k+1] \\ 2k - j + 2, & \text{if } j \in O[k+2, 2k+1] \\ j - 2k - 1, & \text{if } j \in O[2k+3, 3k] \\ j - 2k, & \text{if } j \in O[3k+1, n-1] \\ 2k + j, & \text{if } j \in E[2, k+1] \\ 2k + j - 1, & \text{if } j \in E[k+2, 2k] \\ 6k - j + 2, & \text{if } j \in E[2k+2, 3k] \\ 6k - j + 1, & \text{if } j \in E[3k+1, n]. \end{cases}$$

An equivalent formulation for θ_n, which is easier to use in some situations is

$$\theta_n : \begin{cases} 2i-1 \mapsto 2k-2i+2, & \text{if } 1 \le i \le (k+2)/2 \\ 2k-2i+3 \mapsto 2i-1, & \text{if } 1 \le i \le (k+1)/2 \\ 2k+2i+1 \mapsto 2i, & \text{if } 1 \le i \le (k-1)/2 \\ n-2i+1 \mapsto 2k-2i+1, & \text{if } 1 \le i \le k/2 \\ 2i \mapsto 2k+2i, & \text{if } 1 \le i \le (k+1)/2 \\ 2k-2i+2 \mapsto n-2i+1, & \text{if } 1 \le i \le k/2 \\ 2k+2i \mapsto n-2i+2, & \text{if } 1 \le i \le k/2 \\ n-2i+2 \mapsto 2k+2i-1, & \text{if } 1 \le i \le (k+1)/2. \end{cases}$$

For example, for k odd, θ_n is easily seen to be the cycle

$$\begin{gathered}(2k+1\ \ 1\ \ 2k\ \ n-1\ \ldots\ldots\ \overbrace{2k-2i+3\ \ 2i-1\ \ 2k-2i+2\ \ n-2i+1}^{2\le i\le\frac{k-3}{2}}\ \ldots\ldots \\ \ldots\ldots\ k+4\ \ k-2\ \ k+3\ \ 3k+2\ \ k+2\ \ k\ \ k+1\ \ 3k+1\ \ 3k\ \ k-1 \\ 3k-1\ \ 3k+3\ \ldots\ldots\ \overbrace{2k+2i+1\ \ 2i\ \ 2k+2i\ \ n-2i+2}^{\frac{k-3}{2}\ge i\ge 2}\ \ldots\ldots \\ \ldots\ldots\ 2k+3\ \ 2\ \ 2k+2\ \ n).\end{gathered}$$

Similarly, for k even, θ_n is the cycle

$$\begin{gathered}(2k+1\ \ 1\ \ 2k\ \ n-1\ \ldots\ldots\ \overbrace{2k-2i+3\ \ 2i-1\ \ 2k-2i+2\ \ n-2i+1}^{2\le i\le\frac{k-2}{2}}\ \ldots\ldots \\ \ldots\ldots\ k+3\ \ k-1\ \ k+2\ \ 3k+1\ \ k+1\ \ k\ \ 3k\ \ 3k+2\ \ 3k-1\ \ k-2 \\ 3k-2\ \ 3k+4\ \ldots\ldots\ \overbrace{2k+2i+1\ \ 2i\ \ 2k+2i\ \ n-2i+2}^{\frac{k-4}{2}\ge i\ge 2}\ \ldots\ldots \\ \ldots\ldots\ 2k+3\ \ 2\ \ 2k+2\ \ n).\end{gathered}$$

We also note the following general features of f_{θ_n}:

(1) The map f_{θ_n} has a local minimum at $j=1$.
(2) The map f_{θ_n} is maximodal and has all maximum values above all minimum values.
(3) The map f_{θ_n} has a global minimum at $j=2k+1$.
(4) The map f_{θ_n} has a global maximum at $j=2k+2$.

The general "shape" of the graph f_{θ_n} for $n \equiv 0 \pmod 4$ is illustrated in Figures 1 and 2 below.

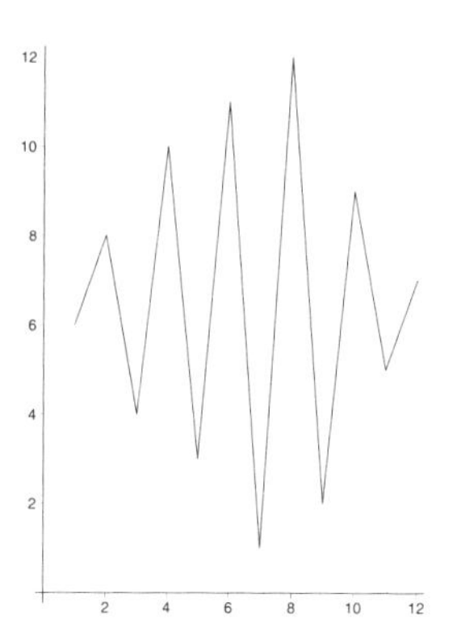

Figure 1. Graph of f_{θ_n} when $n = 12$.

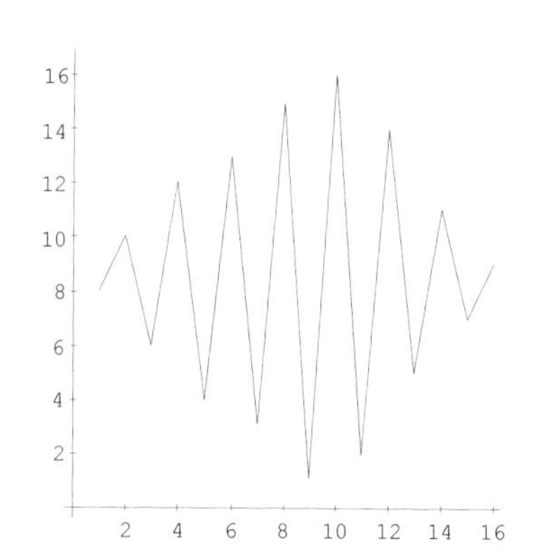

Figure 2. Graph of f_{θ_n} when $n = 16$.

The permutation θ_n is not self dual, however the dual, $\overline{\theta_n}$, is automatically a cycle as it is conjugate to θ_n by φ. In general, the reverse permutation $\widetilde{\theta}$ of a cycle θ is not a cycle, but in the case of θ_n, $\widetilde{\theta_n}$ is a cycle and hence so is θ^*. Specifically, $\widetilde{\theta_n}$ is the cycle

$$\begin{aligned}(2k\ \ 1\ \ 2k+1\ \ n-1\ \ldots\ldots\ &\overbrace{2k+2i-2\ \ 2i-1\ \ 2k+2i-1\ \ n-2i+1}^{2\le i\le \frac{k-3}{2}}\ \ldots\ldots\\ \ldots\ldots\ 3k-3\ \ k-2\ \ 3k-2\ \ &3k+2\ \ 3k-1\ \ k\ \ 3k\ \ 3k+1\ \ k+1\ \ k-1\\ k+2\ \ 3k+3\ \ldots\ldots\ &\overbrace{2k-2i\ \ 2i\ \ 2k-2i+1\ \ n-2i+2}^{\frac{k-3}{2}\ge i\ge 2}\ \ldots\ldots\\ \ldots\ldots\ &2k-2\ \ 2\ \ 2k-1\ \ n)\end{aligned}$$

for k odd, and for k even, $\widetilde{\theta_n}$ is the cycle

$$\begin{aligned}(2k\ \ 1\ \ 2k+1\ \ n-1\ \ldots\ldots\ &\overbrace{2k+2i-2\ \ 2i-1\ \ 2k+2i-1\ \ n-2i+1}^{2\le i\le \frac{k-2}{2}}\ \ldots\ldots\\ \ldots\ldots\ 3k-2\ \ k-1\ \ 3k-1\ \ &3k+1\ \ 3k\ \ k\ \ k+1\ \ 3k+2\ \ k+2\ \ k-2\\ k+3\ \ 3k+4\ \ldots\ldots\ &\overbrace{2k-2i\ \ 2i\ \ 2k-2i+1\ \ n-2i+2}^{\frac{k-4}{2}\ge i\ge 2}\ \ldots\ldots\\ \ldots\ldots\ &2k-2\ \ 2\ \ 2k-1\ \ n).\end{aligned}$$

We now present the main theorem of the paper.

THEOREM 4.2. *For $n \equiv 0 \pmod 4$ the permutations θ_n, $\widetilde{\theta_n}$ and their duals are cycles and have maximum entropy amongst all cycles of period n.*

In the proof we will show that the cycle θ_n has maximum entropy and that $h(\theta_n) = h(\widetilde{\theta_n})$. It then follows that so do $\overline{\theta_n}$ and the dual of $\widetilde{\theta_n}$ (see comments after Proposition 2.3).

We will now consider the induced matrix $B = M(\theta_n)$ of θ_n. The rows of B are given by the following formulae:

PROPOSITION 4.3. *Let $n = 4k$ and $B = M(\theta_n)$. If k is odd then*

$$B_{j\,r} = 1 \Leftrightarrow \begin{cases} 2k - j + 1 \leq r \leq 2k + j, & \text{if } j \in O[1, k] \\ 2k - j + 2 \leq r \leq 2k + j - 1, & \text{if } j \in O[k + 2, 2k - 1] \\ 1 \leq r \leq n - 1, & \text{if } j = 2k + 1 \\ j - 2k - 1 \leq r \leq 6k - j, & \text{if } j \in O[2k + 3, 3k - 2] \\ k - 1 \leq r \leq 3k - 1, & \text{if } j = 3k \\ j - 2k \leq r \leq 6k - j - 1, & \text{if } j \in O[3k + 2, n - 1] \\ 2k - j \leq r \leq 2k + j - 1, & \text{if } j \in E[2, k - 1] \\ k \leq r \leq 3k, & \text{if } j = k + 1 \\ 2k - j + 1 \leq r \leq 2k + j - 2, & \text{if } j \in E[k + 3, 2k] \\ j - 2k \leq r \leq 6k - j + 1, & \text{if } j \in E[2k + 2, 3k - 1] \\ j - 2k + 1 \leq r \leq 6k - j, & \text{if } j \in E[3k + 1, n - 2]. \end{cases}$$

If k is even then

$$B_{j\,r} = 1 \Leftrightarrow \begin{cases} 2k - j + 1 \leq r \leq 2k + j, & \text{if } j \in O[1, k - 1] \\ k \leq r \leq 3k, & \text{if } j = k + 1 \\ 2k - j + 2 \leq r \leq 2k + j - 1, & \text{if } j \in O[k + 3, 2k - 1] \\ 1 \leq r \leq n - 1, & \text{if } j = 2k + 1 \\ j - 2k - 1 \leq r \leq 6k - j, & \text{if } j \in O[2k + 3, 3k - 1] \\ j - 2k \leq r \leq 6k - j - 1, & \text{if } j \in O[3k + 1, n - 1] \\ 2k - j \leq r \leq 2k + j - 1, & \text{if } j \in E[2, k] \\ 2k - j + 1 \leq r \leq 2k + j - 2, & \text{if } j \in E[k + 2, 2k] \\ j - 2k \leq r \leq 6k - j + 1, & \text{if } j \in E[2k + 2, 3k - 2] \\ k + 1 \leq r \leq 3k + 1, & \text{if } j = 3k \\ j - 2k + 1 \leq r \leq 6k - j, & \text{if } j \in E[3k + 2, n - 2]. \end{cases}$$

We can now use these formulae to convert information about rows of B to information about columns of B.

DEFINITION 4.4. Given $a, b \in \mathbb{N}$ with $1 \leq a \leq b \leq n - 1$, we define $\langle a, b \rangle$ to be an $(n-1) \times 1$ column matrix with *ith* row entry $\langle a, b \rangle_i$ given by

$$\langle a, b \rangle_i = \begin{cases} 1, & \text{if } a \leq i \leq b \\ 0, & \text{otherwise.} \end{cases}$$

If column j of matrix M is equal to $\langle a, b \rangle$ then we can write $M^{(j)} = \langle a, b \rangle$.

Note that $\|M^{(j)}\| = |M^{(j)}|$ (as in Notation 2.5) and if $M^{(j)} = \langle a, b \rangle$ then $\|M^{(j)}\| = b - a + 1$.

Furthermore, for any $j \in [1, n-1]$, we define $\langle M^{(j)} \rangle = \{i \in [1, n-1] : m_{i\,j} = 1\}$ and call this set the *column support* of $M^{(j)}$.

PROPOSITION 4.5. *For $n = 4k$ the column $B^{(j)}$ is given by*

$$\begin{array}{ll}
\langle 2k-j+1, 2k+j\rangle, & \text{if } j \in O[1, 2k-1] \\
\langle j-2k, 6k-j-1\rangle, & \text{if } j \in O[2k+1, 3k] \\
\langle j-2k+2, 6k-j+1\rangle, & \text{if } j \in O[3k+1, n-1] \\
\langle 2k-j+2, 2k+j+1\rangle, & \text{if } j \in E[2, k-1] \\
\langle 2k-j, 2k+j-1\rangle, & \text{if } j \in E[k, 2k-2] \\
\langle 1, n-1\rangle, & \text{if } j = 2k \\
\langle j-2k+1, 6k-j\rangle, & \text{if } j \in E[2k+2, n-2].
\end{array}$$

An equivalent formulation of Proposition 4.5 which is needed in our main proofs is

PROPOSITION 4.6. *For $n = 4k$ the column $B^{(j)}$ is given by*

$$\begin{array}{ll}
\langle 2k-2i+2, 2k+2i-1\rangle, & \text{if } j = 2i-1 \text{ and } 1 \le i \le k/2 \\
\langle 2k-2i+2, 2k+2i+1\rangle, & \text{if } j = 2i \text{ and } 1 \le i \le (k-1)/2 \\
\langle 2i, n-2i-1\rangle, & \text{if } j = 2k-2i \text{ and } 1 \le i \le k/2 \\
\langle 2i, n-2i+1\rangle, & \text{if } j = 2k-2i+1 \text{ and } 1 \le i \le (k+1)/2 \\
\langle 1, n-1\rangle, & \text{if } j = 2k \\
\langle 2i-1, n-2i\rangle, & \text{if } j = 2k+2i-1 \text{ and } 1 \le i \le (k+1)/2 \\
\langle 2i+1, n-2i\rangle, & \text{if } j = 2k+2i \text{ and } 1 \le i \le k/2 \\
\langle 2k-2i+1, 2k+2i\rangle, & \text{if } j = n-2i \text{ and } 1 \le i \le (k-1)/2 \\
\langle 2k-2i+3, 2k+2i\rangle, & \text{if } j = n-2i+1 \text{ and } 1 \le i \le k/2.
\end{array}$$

From Proposition 4.5 we can easily obtain

COROLLARY 4.7. *The matrix B has jth column sum*

$$\left|B^{(j)}\right| = \begin{cases} 2j, & \text{if } j < 2k \\ n-1, & \text{if } j = 2k \\ 2(n-j), & \text{if } j > 2k. \end{cases}$$

Corollary 4.7 in conjunction with Corollary 3.3 shows that the matrix B has the maximum column sum for every column of B. It is interesting to note that for $n = 4k$ both the maximum entropy n-permutations (see King [**8**]) and maximum entropy n-cycles are maximodal, have all maximum values above all minimum values and

$$\begin{array}{l}
\theta(2) < \theta(4) < \cdots < \theta(j) > \theta(j+2) > \cdots > \theta(n) \text{ and} \\
\theta(1) > \theta(3) > \cdots > \theta(i) < \theta(i+2) < \cdots < \theta(n-1),
\end{array}$$

where $\theta(j)$ is the global maximum and $\theta(i)$ is the global minimum. This is reflected in the induced matrices which have the property that each column contains the maximum number of 1's and there are no 0's between 1's in any column.

Recall that our task is to demonstrate that the cycles θ_n and $\widetilde{\theta_n}$ and their duals all achieve maximum entropy amongst all cycles of period n. As we know, for **any** permutation θ, $h(\theta)$ is given by log $(\lim_{p\to\infty} \|M(\theta)^p\|^{1/p})$. It follows that if θ and ϕ are permutations, a sufficient but not necessary condition for $h(\theta)$ to be greater than or equal to $h(\phi)$ is that $\|M(\theta)^p\| \geq \|M(\phi)^p\|$ for all $p \in \mathbb{N}$. Given this, we will prove that $\|B^p\| \geq \|C^p\|$ for all $p \in \mathbb{N}$, where $B = M(\theta_n)$ and $C \in \Gamma$, for Γ a class of matrices which we will define. Since the definition of Γ is vital to the proof of our main theorem, we discuss it in some detail. To facilitate this discussion we introduce two simple definitions.

DEFINITION 4.8. A matrix is called a *0-1 matrix* if its only entries are 0 and 1.

DEFINITION 4.9. If X and Y are any two $(n-1)\times(n-1)$, $0-1$ matrices, X is said to *dominate* Y if and only if $\|X^p\| \geq \|Y^p\|$ for all $p \in \mathbb{N}$.

DEFINITION 4.10. The permutation ψ_n defined by:

$$\psi_n : \quad j \mapsto \begin{cases} 2k-j+1, & \text{if } j \in O[1, 2k-1] \\ j-2k, & \text{if } j \in O[2k+1, n-1] \\ 2k+j, & \text{if } j \in E[2, 2k] \\ 6k-j+1, & \text{if } j \in E[2k+2, n]. \end{cases}$$

is the unique n-permutation which has maximum entropy in P_n and for which f_{ψ_n} has a local minimum at 1 (see King [**8, 9**]).

NOTATION 4.11. The symbol A is reserved to denote the induced matrix of ψ_n.

Recall, we have shown that we need only consider n-cycles which are maximodal. Furthermore, we will only consider n-cycles ϕ for which f_ϕ has a local minimum at 1 (the case where f_ϕ has a local maximum at 1 will be dealt with later). Our tactic is to define a class Γ of $(n-1)\times(n-1)$, $0-1$ matrices with the following properties:

1. For all $C \in \Gamma$, B dominates C.
2. For all n-cycles ϕ for which f_ϕ is maximodal and has a local minimum at 1, there is a C in Γ such that C dominates $M(\phi)$.

As dominance is clearly a transitive relationship, this will suffice to show that $h(\theta_n) \geq h(\phi)$ for any n-cycle ϕ for which f_ϕ is maximodal and has a local minimum at 1. In a later section we demonstrate that θ_n^*, the dual of the reverse of θ_n, is a cycle of identical topological entropy to θ_n and has the property that $h(\theta_n^*) \geq h(\phi)$ for any n-cycle ϕ for which f_ϕ is maximodal and has a local maximum at 1. As we only need to consider cycles which are maximodal, this will complete the proof of Theorem 4.2.

It turns out that $B \in \Gamma$ (a likely but not necessary consequence of properties 1 and 2) and in view of property 2, Γ need not contain the induced matrices of all maximodal n-cycles for which f_ϕ has a local minimum at 1.

It is clear that we cannot permit A to be an element of Γ but the induced matrices of other n-permutations may be elements of Γ. In fact we only need to exclude the induced matrices of a relatively small class of these non-cyclic n-permutations from Γ. Those to be excluded arise from Proposition 4.12.

PROPOSITION 4.12. *Let D be the induced matrix of an n-permutation ϕ such that ϕ is maximodal with $\phi(1) < \phi(2)$ and*

1. *For some $i \in [1,(k-1)/2]$ the matrix D is identical to the matrix A on each of the four columns $2i,\ 2k-2i,\ 2k+2i,\ n-2i$*

   or

2. *For some $i \in [1,k/2]$ the matrix D is identical to the matrix A on each of the four columns $2i-1,\ 2k-2i+1,\ 2k+2i-1,\ n-2i+1$.*

Then ϕ is not a cycle.

PROOF. Clearly it suffices to show that a proper subset of $[1,n]$ is fully invariant under ϕ to demonstrate that ϕ is not a cycle. If D satisfies condition 1 for a specific $i \in [1,(k-1)/2]$ then the set $[2i+1,2k-2i] \cup [2k+2i+1,n-2i]$ is a proper subset of $[1,n]$ and is fully invariant under ϕ. The former is trivial, since for $i \in [1,(k-1)/2]$, $[2i+1,2k-2i] \cup [2k+2i+1,n-2i] \subseteq [3,2k-2] \cup [2k+3,n-2]$. The latter may easily be established in a case by case fashion using the method for retrieving ϕ from D suggested in the comments following Lemma 3.2. To see how this works in a typical case, suppose for example that j is an odd element of $[2i+1,2k-2i] \cup [2k+2i+1,n-2i]$. Trivially, $j \in O[2i+1,2k-2i-1] \cup O[2k+2i+1,n-2i-1]$. We wish to show that $\phi(j) \in [2i+1,2k-2i] \cup [2k+2i+1,n-2i]$. Recall that the method of retrieving $\phi(j)$ from D, where j is odd, involves scanning row j of D from the left and stopping at the first 1. The number $\phi(j)$ is then the number of the column containing this 1. Note that since $j \in [2i+1,2k-2i-1] \cup [2k+2i+1,n-2i-1]$, j is not in the column support $[2k-2i,2k+2i-1]$ of $D^{(2i)}$, but is in the column support $[2i,n-2i-1]$ of $D^{(2k-2i)}$. In particular, this means $d_{j\,2i} = 0$ and $d_{j\,2k-2i} = 1$ and hence as there are no "gaps" in rows of 1's in D, $2i < \phi(j) \leq 2k-2i$. Thus $\phi(j) \in [2i+1,2k-2i] \subseteq [2i+1,2k-2i] \cup [2k+2i+1,n-2i]$ as required.

The case where j is even follows similarly with $D^{(2k+2i)} = \langle 2i+1,n-2i \rangle$ and $D^{(n-2i)} = \langle 2k-2i+1,2k+2i \rangle$ being used to show $\phi(j) \in [2k+2i+1,n-2i]$. This completes the cases where D satisfies condition 1 for a specific $i \in [1,(k-1)/2]$.

If the matrix D satisfies condition 2 for a specific $i \in [1,k/2]$ then the set $S = [2i,2k-2i+1] \cup [2k+2i,n-2i+1]$ is a proper subset of $[1,n]$ and can be similarly shown to be fully invariant under ϕ by showing that if j is an odd element of S then $\phi(j) \in [2i,2k-2i+1]$ and if j is an even element of S then $\phi(j) \in [2k+2i,n-2i+1]$. □

Proposition 4.12 identified non-cyclic n-permutations which have a fully invariant proper subset of a particular type. We will exclude the induced matrix of any such n-permutation from Γ. However, we will accept the induced matrix C of any maximodal n-permutation ϕ with $\phi(1) < \phi(2)$ and for which $C^{(2k)} = \langle 1,n-1 \rangle$ as an element of Γ provided C does not satisfy the hypotheses of Proposition 4.12. That is, provided

(i) For all $i \in [1,(k-1)/2]$ **at least one** of the following inequalities holds:

$$C^{(2i)} \neq \langle 2k-2i,2k+2i-1 \rangle \qquad \left(= A^{(2i)} \neq B^{(2i)}\right)$$

$$C^{(2k-2i)} \neq \langle 2i,n-2i-1 \rangle \qquad \left(= A^{(2k-2i)} = B^{(2k-2i)}\right)$$

$$C^{(2k+2i)} \neq \langle 2i+1,n-2i \rangle \qquad \left(= A^{(2k+2i)} = B^{(2k+2i)}\right)$$

$$C^{(n-2i)} \neq \langle 2k-2i+1,2k+2i \rangle \qquad \left(= A^{(n-2i)} = B^{(n-2i)}\right),$$

(ii) For all $i \in [1, k/2]$ **at least one** of the following inequalities holds:

$$\begin{aligned} C^{(2i-1)} &\neq \langle 2k-2i+2, 2k+2i-1\rangle & &\left(= A^{(2i-1)} = B^{(2i-1)}\right) \\ C^{(2k-2i+1)} &\neq \langle 2i, n-2i+1\rangle & &\left(= A^{(2k-2i+1)} = B^{(2k-2i+1)}\right) \\ C^{(2k+2i-1)} &\neq \langle 2i-1, n-2i\rangle & &\left(= A^{(2k+2i-1)} = B^{(2k+2i-1)}\right) \\ C^{(n-2i+1)} &\neq \langle 2k-2i+1, 2k+2i-2\rangle & &\left(= A^{(n-2i+1)} \neq B^{(n-2i+1)}\right). \end{aligned}$$

This makes it evident that B belongs to Γ but A does not.

The maximodal n-cycles with $\phi(1) < \phi(2)$ of greatest concern to us are those whose induced matrices have columns which match A on columns where $B^{(j)} \neq A^{(j)}$. To help us handle these cases we have the following result:

COROLLARY 4.13. *If D is the induced matrix of a maximodal n-cycle with $\phi(1) < \phi(2)$ such that*

1. *For some $i \in [1, (k-1)/2]$, $D^{(2i)} = A^{(2i)} \neq B^{(2i)}$*

   or

2. *For some $i \in [1, k/2]$, $D^{(n-2i+1)} = A^{(n-2i+1)} \neq B^{(n-2i+1)}$,*

then there exists $j' \in \{2k-2i, 2k+2i, n-2i\}$ or $j' \in \{2k-2i+1, 2k+2i-1, 2i-1\}$ in cases 1 and 2 respectively, such that $D^{(j')} \neq A^{(j')} = B^{(j')}$.

The class Γ of $(n-1) \times (n-1)$, $0-1$ matrices we will consider is defined as follows:

DEFINITION 4.14. The $(n-1) \times (n-1)$, $0-1$ matrix $C \in \Gamma$ if and only if

1. $C^{(2k)} = \langle 1, n-1\rangle$.

   (So if C is the induced matrix of a maximodal permutation ϕ, all maximum values of ϕ are above all minimum values of ϕ.)

2. If for some i and $j_1 \leq j_2$ we have $c_{i\,j_1} = c_{i\,j_2} = 1$ then $c_{i\,j} = 1$ for all $j \in [j_1, j_2]$.

   (So C has no row gaps.)

3. If $c_{i-1\,j} = c_{i\,j} = 1$ for some j and some odd (respectively, even) i, then $c_{i-1\,j'} = c_{i\,j'}$ for all $j' \leq j$ (respectively, $j' \geq j$).

   (So C satisfies the Shape Lemma appropriate to induced matrices of maximodal permutations ϕ for which $\phi(1) < \phi(2)$. Note that in view of (1) we can conclude that for $i \in O[3, n-1]$, $c_{i-1\,j'} = c_{i\,j'}$ for all $j' \leq 2k$ and for $i \in E[2, n-2]$, $c_{i-1\,j'} = c_{i\,j'}$ for all $j' \geq 2k$.)

4. For all $j \neq 2k$

$$\left|C^{(j)}\right| \leq \left|B^{(j)}\right| = \left|A^{(j)}\right| = \min\,\{2j, 2(n-j)\}.$$

   (So C has the necessary permutation shape.)

5.
(i) For all $i \in [1,(k-1)/2]$ **at least one** of the following inequalities holds:

$$\begin{aligned} C^{(2i)} &\neq A^{(2i)} && (= \langle 2k-2i, 2k+2i-1\rangle), \\ C^{(2k-2i)} &\neq A^{(2k-2i)} && (= \langle 2i, n-2i-1\rangle), \\ C^{(2k+2i)} &\neq A^{(2k+2i)} && (= \langle 2i+1, n-2i\rangle), \\ C^{(n-2i)} &\neq A^{(n-2i)} && (= \langle 2k-2i+1, 2k+2i\rangle), \end{aligned}$$

(ii) For all $i \in [1,k/2]$ **at least one** of the following inequalites holds:

$$\begin{aligned} C^{(2i-1)} &\neq A^{(2i-1)} && (= \langle 2k-2i+2, 2k+2i-1\rangle), \\ C^{(2k-2i+1)} &\neq A^{(2k-2i+1)} && (= \langle 2i, n-2i+1\rangle), \\ C^{(2k+2i-1)} &\neq A^{(2k+2i-1)} && (= \langle 2i-1, n-2i\rangle), \\ C^{(n-2i+1)} &\neq A^{(n-2i+1)} && (= \langle 2k-2i+1, 2k+2i-2\rangle). \end{aligned}$$

(So C satisfies the cycle condition.)

REMARK 4.15. Observe that B satisfies conditions 1–5 in the definition of Γ so $B \in \Gamma$ whilst A satisfies conditions 1–4 but does not satisfy condition 5 so $A \notin \Gamma$. In particular, the induced matrix $M(\phi)$ of any maximodal n-permutation ϕ with $\phi(1) < \phi(2)$ and which has all maximum values above all minimum values satisfies conditions 1–4 of Γ. So $M(\phi) \in \Gamma$ if and only if it satisfies condition 5. A consequence of our proof that B dominates each element C of Γ will thus be that $h(\theta_n) \geq h(\phi)$ if $M(\phi)$ satisfies condition 5 (where ϕ is maximodal, $\phi(1) < \phi(2)$ and all maximum values of ϕ are above all minimum values of ϕ). Notice that Γ does not contain all possible induced matrices arising from maximodal n-cycles ϕ with $\phi(1) < \phi(2)$. This is because the induced matrix of such an n-cycle which also has at least one maximum value less than at least one minimum value will fail condition 1 of Definition 4.14. It will be seen in Lemma 5.4 in the next section that the maximodal n-cycles ϕ for which $\phi(1) < \phi(2)$ that are not elements of Γ are precisely those with at least one maximum value less than at least one minimum value. However this causes no problem since every such matrix is dominated by another element of Γ.

5. Some easy lemmas

Throughout this section we will be dealing with $(n-1)\times(n-1)$, $0-1$ matrices. Our aim is to present results which are helpful in the calculation of $\|M^p\|$ or in comparing $\|M^p\|$ with $\|N^p\|$ for any $p \in \mathbb{N}$. Lemma 5.5 and Corollaries 5.6, 5.7 and 5.13 are of particular importance.

NOTATION 5.1. Let M be an $(n-1)\times(n-1)$, $0-1$ matrix.

1. For any $T \subseteq [1, n-1]$ and any $p \in \mathbb{N}$

$$\sum_{j \in T} \left|(M^p)^{(j)}\right| \quad \text{will be denoted by } M^p_T.$$

   Thus, for example, $\|M^p\| = M^p_{[1,n-1]}$.
2. The set $\{i \in [1, n-1] : m_{i\,j} = 1\}$ will be denoted by $S_j(M)$.

This latter notation will be of particular use in inductive arguments because of the next lemma.

LEMMA 5.2. *Let M be an $(n-1)\times(n-1)$, $0-1$ matrix, then*

$$\left|\left(M^{p+1}\right)^{(j)}\right| = \sum_{i\in S_j(M)} \left|\left(M^{p}\right)^{(i)}\right| = M^p_{S_j(M)}$$

for all $p \in \mathbb{N}$ and $j \in [1, n-1]$.

Note that if we interpret M^0 as the $(n-1)\times(n-1)$ identity matrix, then Lemma 5.2 remains valid for $p = 0$.

PROOF. Notice that if X is any $(n-1)\times(n-1)$ matrix and $\mathbf{v}$ is the $1\times(n-1)$ matrix $(1,1,1,\dots,1)$ then $\mathbf{v}X$ is the $1\times(n-1)$ matrix $\left(\left|X^{(1)}\right|, \left|X^{(2)}\right|, \dots, \left|X^{(n-1)}\right|\right)$. Thus

$$\begin{aligned}\left(\left|\left(M^{p+1}\right)^{(1)}\right|, \left|\left(M^{p+1}\right)^{(2)}\right|, \dots, \left|\left(M^{p+1}\right)^{(n-1)}\right|\right) \\ &= \mathbf{v}M^{p+1} \\ &= \left(\mathbf{v}M^{p}\right) M \\ &= \left(\left|\left(M^{p}\right)^{(1)}\right|, \left|\left(M^{p}\right)^{(2)}\right|, \dots, \left|\left(M^{p}\right)^{(n-1)}\right|\right) M.\end{aligned}$$

Thus

$$\begin{aligned}\left|\left(M^{p+1}\right)^{(j)}\right| &= \sum_{i=1}^{n-1} \left|\left(M^{p}\right)^{(i)}\right| m_{i\,j} \\ &= \sum_{i\in S_j(M)} \left|\left(M^{p}\right)^{(i)}\right|,\end{aligned}$$

since

$$m_{i\,j} = \begin{cases} 1, & \text{if } i \in S_j(M) \\ 0, & \text{if } i \in [1, n-1] \setminus S_j(M). \end{cases}$$

□

COROLLARY 5.3. *Let M and N be $(n-1)\times(n-1)$, $0-1$ matrices. A sufficient condition for N to dominate M is that $S_j(M) \subseteq S_j(N)$ for all $j \in [1, n-1]$; (that is, that $m_{i\,j} = 1 \implies n_{i\,j} = 1$ for all $i, j \in [1, n-1]$).*

PROOF. It is easy to see by Lemma 5.2 and induction on p that $\left|\left(M^{p}\right)^{(j)}\right| \leq \left|\left(N^{p}\right)^{(j)}\right|$ and hence that $\|M^p\| \leq \|N^p\|$ for all $p \in \mathbb{N}$. □

We are now in a position to substantiate our claim at the end of the previous section.

LEMMA 5.4. *Let ϕ be a maximodal n-cycle for which $\phi(1) < \phi(2)$ and let $M(\phi)$ be its induced matrix. Then there exists an element $C \in \Gamma$ such that C dominates $M(\phi)$.*

PROOF. We know $M(\phi)$ has no row gaps and satisfies the Shape Lemma; that is, $M(\phi)$ satisfies conditions 2 and 3 in the definition of Γ. By Proposition 3.5, $M(\phi)$ satisfies condition 4 in the definition of Γ, and by Proposition 4.12, since ϕ is a cycle, $M(\phi)$ satisfies condition 5 in the definition of Γ. Therefore $M(\phi)$ is in Γ if and only if $M(\phi)$ also satisfies condition 1. Clearly, if $M(\phi)^{(2k)} = \langle 1, n-1 \rangle$ then $M(\phi) \in \Gamma$ and setting $C = M(\phi)$ does the job. So assume $M(\phi)^{(2k)} \neq \langle 1, n-1 \rangle$.

By Remark 3.1 and as ϕ is maximodal with $\phi(1) < \phi(2)$, $M(\phi)^{(2k)} = \langle 1, n-1 \rangle$ if and only if $\phi(i) \leq 2k$ for all $i \in O[1, n-1]$ and $\phi(i) > 2k$ for all $i \in E[2, n]$. This means that all maximum values of f_ϕ are above all minimum values of f_ϕ. Since we have assumed that $M(\phi)^{(2k)} \neq \langle 1, n-1 \rangle$ then there is at least one maximum value below one minimum value; that is, $\phi(i) \leq 2k$ for some $i \in E[2, n]$ and $\phi(i) > 2k$ for some $i \in O[1, n-1]$.

So let $P = \phi^{-1}([2k+1, n]) \cap O[1, n-1] \neq \emptyset$ and $Q = \phi^{-1}([1, 2k]) \cap E[2, n] \neq \emptyset$ and note that P and Q have the same cardinality since ϕ is a bijection.

Given these preliminaries and making use of the information on $M(\phi)$ in Remark 3.1, we can now define the $(n-1) \times (n-1)$, $0-1$ matrix C chosen to be a member of Γ and to dominate $M(\phi)$ as follows:

(i) If $p \in P$; that is, $p \in O[1, n-1]$ and $\phi(p) \geq 2k+1$,

$$c_{pj} = \begin{cases} 1, & \text{if } j \in [2k, \phi(p+1)-1] \\ 0, & \text{if } j \notin [2k, \phi(p+1)-1] \end{cases}$$

while

$$M(\phi)_{pj} = \begin{cases} 1, & \text{if } j \in [\phi(p), \phi(p+1)-1] \subset [2k, \phi(p+1)-1] \\ 0, & \text{if } j \notin [\phi(p), \phi(p+1)-1]. \end{cases}$$

If in addition, $p \neq 1$,

$$c_{p-1\,j} = \begin{cases} 1, & \text{if } j \in [2k, \phi(p-1)-1] \\ 0, & \text{if } j \notin [2k, \phi(p-1)-1] \end{cases}$$

while

$$M(\phi)_{p-1\,j} = \begin{cases} 1, & \text{if } j \in [\phi(p), \phi(p-1)-1] \subset [2k, \phi(p-1)-1] \\ 0, & \text{if } j \notin [\phi(p), \phi(p-1)-1]. \end{cases}$$

Note the implication here that if $p \in P$, $p+1 \notin Q$ and further if $p \in P$ and $p \neq 1$, $p-1 \notin Q$.

(ii) If $q \in Q$; that is, $q \in E[2, n]$ and $\phi(q) \leq 2k$,

$$c_{q-1\,j} = \begin{cases} 1, & \text{if } j \in [\phi(q-1), 2k] \\ 0, & \text{if } j \notin [\phi(q-1), 2k] \end{cases}$$

while

$$M(\phi)_{q-1\,j} = \begin{cases} 1, & \text{if } j \in [\phi(q-1), \phi(q)-1] \subset [\phi(q-1), 2k] \\ 0, & \text{if } j \notin [\phi(q-1), \phi(q)-1]. \end{cases}$$

If in addition, $q \neq n$,

$$c_{qj} = \begin{cases} 1, & \text{if } j \in [\phi(q+1), 2k] \\ 0, & \text{if } j \notin [\phi(q+1), 2k] \end{cases}$$

while

$$M(\phi)_{qj} = \begin{cases} 1, & \text{if } j \in [\phi(q+1), \phi(q)-1] \subset [\phi(q+1), 2k] \\ 0, & \text{if } j \notin [\phi(q+1), \phi(q)-1]. \end{cases}$$

Note the implication here that if $q \in Q$, $q-1 \notin P$ and further if $q \in Q$ and $q \neq n$, $q+1 \notin P$.

(iii) If $i \in [1, n-1]$ and c_{ij} has not been specified in (i) or (ii) (that is, $\{i, i+1\} \cap (P \cup Q) = \emptyset$) then

$$c_{ij} = M(\phi)_{ij}.$$

This may be more simply expressed without emphasizing the structure of $M(\phi)$ by noting that the modification made to $M(\phi)$ to produce C is to replace any 0 in the $2k\,th$ column by 1 and then to replace the minimum number of 0's remaining by 1's in order to ensure that there are no row gaps.

It is evident by this construction that C dominates $M(\phi)$ and that C satisfies conditions 1, 2 and 3 in the definition of Γ.

We now show that C satisfies condition 5. In view of the fact that $M(\phi)$ satisfies condition 5, it suffices to show for each $j \in [1, n-1]$ that

$$C^{(j)} = A^{(j)} \implies \left(M(\phi)^{(j)} = A^{(j)} \text{ or } C^{(n-j)} \neq A^{(n-j)}\right)$$

or equivalently, that

$$\left(C^{(j)} = A^{(j)} \text{ and } M(\phi)^{(j)} \neq A^{(j)}\right) \implies C^{(n-j)} \neq A^{(n-j)}.$$

We will consider the case where $j < 2k$. The case where $j > 2k$ can be proved in similar fashion.

Thus let $C^{(j)} = A^{(j)} \neq M(\phi)^{(j)}$ and $j < 2k$. As $C^{(j)} \neq M(\phi)^{(j)}$ and $j < 2k$, the definition of C, condition (ii) implies the existence of $q \in Q$ such that $\phi(q) - 1 < j$ with $c_{q-1\,j} = 1$ and $m_{q-1\,j} = 0$. Further, since $n-j > 2k$, $c_{q-1\,n-j} = 0$. In addition to this, if $q \neq n$ then $c_{q\,j} = 1$ and $c_{q\,n-j} = 0$. As $C^{(j)} = A^{(j)} = \langle x, y\rangle$ say, we have $x \leq q-1 \leq y$ and if $q \neq n$, $x \leq q \leq y$. Thus $x+1 \leq q \leq y+1$ and if $q \neq n$ $x-1 \leq q-1 \leq y-1$. Since

$$A^{(n-j)} = \begin{cases} \langle x+1, y+1\rangle, & \text{if } j \text{ is even} \\ \langle x-1, y-1\rangle, & \text{if } j \text{ is odd} \end{cases}$$

and if $q \neq n$ both $c_{q\,n-j} = 0$ and $c_{q-1\,n-j} = 0$ and we see $C^{(n-j)} \neq A^{(n-j)}$ as required.

This covers the case where $q \neq n$. We show the case $q = n$ does not arise as it leads to a contradiction.

If $q = n$ we still have $c_{n-1\,j} = 1$, $c_{n-1\,n-j} = 0$ and $x \leq n-1 \leq y$. This latter inequality implies in turn that $y = n-1$, $x = 2$ and $j = 2k-1$. Thus $A^{(n-j)} = A^{(2k+1)} = \langle 1, n-2\rangle$. Note that since $Q \neq \emptyset$, $P \neq \emptyset$; that is, there exists $p \in O[1, n-1]$ such that $\phi(p) > 2k$, so as $C^{(2k-1)} = A^{(2k-1)} = \langle 2, n-1\rangle$, there is exactly one $p \in O[1, n-1]$ such that $\phi(p) > 2k$ and indeed $p = 1$. Combining this with the fact that $\phi(n-1) < \phi(n) < 2k$ we conclude that there exists $p' \in$

$O[3, n-3]$ such that $\phi(p') = 2k$. But then condition (iii) in the definition of C gives $c_{p'\,2k-1} = M(\phi)_{p'\,2k-1} = 0$ and $p' \in [2, n-1]$. Hence the contradiction and so $C^{(2k-1)} \neq A^{(2k-1)}$.

It remains to show that C satisfies condition 4.

Remember we have assumed that $M(\phi)^{(2k)} \neq \langle 1, n-1 \rangle$ and hence P and Q are sets of the same non-zero cardinality. Let the cardinality of P and Q be $m \geq 1$ and let $P = \{p_1, p_2, \ldots, p_m\}$ and $Q = \{q_1, q_2, \ldots, q_m\}$ with the elements of P and Q listed in any order desired. For the specific listing chosen, define a new permutation $\psi : [1, n] \to [1, n]$ as follows:

$$\begin{aligned}
\psi(j) &:= \phi(j), \text{ if } j \in [1, n] \setminus (P \cup Q) \\
\psi(p_i) &:= \phi(q_i), \text{ if } i = 1, 2, \ldots, m \\
\psi(q_i) &:= \phi(p_i), \text{ if } i = 1, 2, \ldots, m.
\end{aligned}$$

Trivially, ψ is a permutation with f_ψ maximodal, $\psi(1) < \psi(2)$ and all maximum values of f_ψ lie above all minimum values.

The now familiar analysis of $M(\psi)$ based on Remark 3.1 together with the evident facts that for p odd, $\psi(p) \leq \min\{2k, \phi(p)\}$ and for q even, $\psi(q) \geq \max\{2k+1, \phi(q)\}$ show that $M(\psi)$ dominates C, which in turn dominates $M(\phi)$. Thus, specifically for any j,

$$\left|C^{(j)}\right| \leq \left|M(\psi)^{(j)}\right| \leq \min\{2j, 2(n-j)\}$$

by Proposition 3.5. □

Our basic aim now is to show that for any $C \in \Gamma$, $B = M(\theta_n)$ dominates C; that is, $\|B^p\| \geq \|C^p\|$ for all $p \in \mathbb{N}$. We use induction to do this (Lemma 6.4). Although we have some information about the structure of the matrices in Γ, we do not have nearly enough to phrase our inductive hypothesis as simply "$\|B^p\| \geq \|C^p\|$" and so our inductive hypothesis needs considerable strengthening. Clearly, the inductive hypothesis "$\left|(B^p)^{(j)}\right| \geq \left|(C^p)^{(j)}\right|$" for all $j \in [1, n-1]$ would be sufficient if it were true. Unfortunately the result is not true for all $j \in [1, n-1]$ but it is true for most $j \in [1, n-1]$ and this forms a central part of the inductive hypothesis (Lemma 6.4). For those $j \in [1, n-1]$ where the condition fails we add other column sums to $\left|(B^p)^{(j)}\right|$ in a variety of ways, so that the desired inequality is achieved.

The remainder of this section is devoted to building the machinery needed to facilitate the establishment of the desired column sum comparisons. The next few results are aimed at establishing Lemma 5.10.

Lemma 5.5. *For any $0-1$ matrix M, if $S_j(M) \subseteq S_{j'}(M)$ then for all $p \in \mathbb{N} \cup \{0\}$,*

$$\left|(M^p)^{(j)}\right| \leq \left|(M^p)^{(j')}\right|.$$

Corollary 5.6. *If $C \in \Gamma$ and $1 \leq j \leq j' \leq 2k$ or $2k \leq j' \leq j \leq n-1$ then for all $p \in \mathbb{N} \cup \{0\}$,*

$$\left|(C^p)^{(j)}\right| \leq \left|(C^p)^{(j')}\right|.$$

COROLLARY 5.7. *If* $C \in \Gamma$, $S \subseteq [1, n-1]$, $T \subseteq [1, n-1]$ *and there is an injection* $\rho : S \mapsto T$ *with the property that for all* $j \in S$

$$j \leq \rho(j) \leq 2k \text{ or } 2k \leq \rho(j) \leq j,$$

then for all $p \in \mathbb{N} \cup \{0\}$

$$C_S^p \leq C_T^p.$$

NOTATION 5.8. For $j \in [1, n-1]$ and $B = M(\theta_n)$, let

$$X_j = \{i \in [1, n-1] : b_{i\,j} = 1\} = S_j(B)$$

and for any specific $C \in \Gamma$ let

$$Y_j = \{i \in [1, n-1] : c_{i\,j} = 1\} = S_j(C).$$

(Thus if the column $B^{(j)}$ is equal to $\langle x, y \rangle$ then $X_j = [x, y]$ and if the column $C^{(j)}$ is equal to $\langle a, b \rangle$ then $Y_j = [a, b]$. Of course in the case of C, Y_j may not be a consecutive string of natural numbers.)

Observe here that by Lemma 5.2

$$\left|\left(B^{p+1}\right)^{(j)}\right| = B_{X_j}^p = \sum_{i \in X_j} \left|(B^p)^{(i)}\right|$$

and

$$\left|\left(C^{p+1}\right)^{(j)}\right| = C_{Y_j}^p = \sum_{i \in Y_j} \left|(C^p)^{(i)}\right|.$$

It follows that when we try to compare terms of the form $B_{X_j}^p$ and $C_{Y_j}^p$ the index sets X_j and Y_j play a central role. Clearly, if $X_j = Y_j$ then our task is much simpler than if $X_j \neq Y_j$ but this is unlikely to be the case since C is an arbitrary element of Γ. However, if we can find a subset W of $[1, n-1]$ which is in a sense "closer" to X_j than Y_j is to X_j and which satisfies $C_W^p \geq C_{Y_j}^p$, we may find the desired comparison easier to establish. The search for an appropriate W proceeds through Lemma 5.10, Corollary 5.11 and Lemma 5.12 to the general result in Corollary 5.13. Corollary 5.7 acts as the key tool in the proof of these results. To proceed on this path we first need to establish certain restrictions on Y_j in Lemma 5.9. The validity of these restrictions are trivial consequences of the Shape Lemma properties of elements of Γ, (namely if $C \in \Gamma$ then for $i \in O[3, n-1]$, $c_{i-1\,j'} = c_{i\,j'}$ for all $j' \leq 2k$ and for $i \in E[2, n-2]$ $c_{i-1\,j'} = c_{i\,j'}$ for all $j' \geq 2k$).

LEMMA 5.9. *If* $C \in \Gamma$ *and* $j \in [1, n-1]$ *then*

1. *For* $j < 2k$,
   (a) $Y_j \cap [2k, n-1]$ *has an even number of elements, and*
   (b) $Y_j \cap [1, 2k-1]$ *has an even number of elements if and only if* $1 \notin Y_j$.
2. *For* $j > 2k$,
   (a) $Y_j \cap [1, 2k]$ *has an even number of elements, and*
   (b) $Y_j \cap [2k+1, n-1]$ *has an even number of elements if and only if* $n-1 \notin Y_j$.

The structure of an arbitrary column of a matrix $C \in \Gamma$ can vary widely (except of course $C^{(2k)}$). This creates a difficulty when we try to compare $\left|(C^p)^{(j)}\right|$ with $\left|(B^p)^{(j)}\right|$. However we can make this task much simpler by defining sets (depending on j) which give us an upper bound on the size of $\left|(C^p)^{(j)}\right|$. We need then only compare $\left|(B^p)^{(j)}\right|$ to the appropriate upper bounds. The remainder of the results in this section establish these sets.

LEMMA 5.10. *If $C \in \Gamma$ and $j \in [1, n-1]$, then*

(i) *For $j < 2k$ let 2β be the number of elements in $Y_j \cap [2k, n-1]$ and let $V_j = [2k + 2\beta - 2j, 2k + 2\beta - 1] \cap [1, n-1]$, and*

(ii) *For $j > 2k$ let 2β be the number of elements in $Y_j \cap [1, 2k]$ and let $V_j = [2k - 2\beta + 1, 2k - 2\beta + 2(n-j)] \cap [1, n-1]$.*

Then

$$0 \leq \beta \leq \min \{k, j, n-j\}$$

and for all $p \in \mathbb{N} \cup \{0\}$,

$$C^p_{Y_j} \leq C^p_{V_j}.$$

PROOF. Since 2β is bounded above by the number of elements in Y_j, we have $2\beta \leq \min \{2j, 2(n-j)\}$ by condition 4 of Γ. Also 2β is bounded by either the number of elements in $[2k, n-1]$ or the number of elements in $[1, 2k]$, which in either case is $2k$, thus $0 \leq \beta \leq \min \{k, j, n-j\}$. We now give the proof for $j < 2k$. The proof for $j > 2k$ is similar.

Let $j < 2k$. Note that the number of elements in $Y_j \leq 2j$ and that $Y_j \cap [2k, n-1]$ has 2β elements. It follows that the number of elements in $Y_j \cap [1, 2k-1]$ is bounded above by $\max \{2j - 2\beta, 2k - 1\}$. Note that $[2k, 2k + 2\beta - 1]$ has 2β elements and there is an injection $\rho_1 : Y_j \cap [2k, n-1] \to [2k, 2k + 2\beta - 1]$ such that $2k \leq \rho_1(j) \leq j$. Also, $[2k + 2\beta - 2j, 2k - 1] \cap [1, 2k-1]$ has $\max \{2j - 2\beta, 2k - 1\}$ elements and there is an injection $\rho_2 : Y_j \cap [1, 2k-1] \to [2k + 2\beta - 2j, 2k - 1] \cap [1, 2k-1]$ such that $j \leq \rho_2(j) \leq 2k - 1$. Combining these, there is an injection $\rho : Y_j \to V_j$ where $Y_j = (Y_j \cap [1, 2k-1]) \cup (Y_j \cap [2k, n-1])$ and $V_j = ([2k + 2\beta - 2j, 2k - 1] \cap [1, 2k - 1]) \cup [2k, 2k + 2\beta - 1]$, such that for all $j \in Y_j$, $j \leq \rho(j) \leq 2k$ or $2k \leq \rho(j) \leq j$. Thus by Corollary 5.7, $C^p_{Y_j} \leq C^p_{V_j}$ for all $p \in \mathbb{N} \cup \{0\}$. □

Obviously V_j is constructed from Y_j by compressing elements of Y_j towards the centre point $2k$ and by adding an appropriate number of elements so that V_j still satisfies the conditions of Lemma 5.9, has no more elements than X_j and the number of elements in $Y_j \cap [2k, n-1]$ and $V_j \cap [2k, n-1]$ (respectively $Y_j \cap [1, 2k]$ and $V_j \cap [1, 2k]$) are the same for $j < 2k$ (respectively $j > 2k$).

One point to note, which we will expand upon shortly, is that for each $j \neq 2k$ there is an appropriate value for β such that $V_j = S_j(A)$. We wish to emphasize this fact by rephrasing and slightly strengthening Lemma 5.10 into the next corollary.

COROLLARY 5.11. *Let $C \in \Gamma$ and $j \in [1, 2k-1] \cup [2k+1, n-1]$ and let T_j be a collection of sets of consecutive integers defined as follows:*

$$T_j = \begin{cases} \left\{[2k-j+1+2\alpha, 2k+j+2\alpha] : \frac{(-j-1)}{2} \le \alpha \le \frac{(j-1)}{2}\right\}, & \text{if } j \in O[1, k-1] \\ \{[1, 2j-1]\} \cup \left\{[2k-j+1+2\alpha, 2k+j+2\alpha] : \right. \\ \qquad \left. \frac{(j+1-2k)}{2} \le \alpha \le \frac{(2k-j-1)}{2}\right\}, & \text{if } j \in O[k, 2k-1] \\ \{[2k-j+2\alpha, 2k+j-1+2\alpha] : \frac{-j}{2} \le \alpha \le \frac{j}{2}\}, & \text{if } j \in E[2, k-1] \\ \{[1, 2j-1]\} \cup \left\{[2k-j+2\alpha, 2k+j-1+2\alpha] : \frac{(j+2-2k)}{2} \le \alpha \le \frac{(2k-j)}{2}\right\}, \\ & \text{if } j \in E[k, 2k-2] \\ \{[2j-n+1, n-1]\} \cup \left\{[j-2k+2\alpha, 6k-j-1+2\alpha] : \right. \\ \qquad \left. \frac{(1+2k-j)}{2} \le \alpha \le \frac{(j-2k-1)}{2}\right\}, & \text{if } j \in O[2k+1, 3k] \\ \left\{[j-2k+2\alpha, 6k-j-1+2\alpha] : \frac{(j+1-n)}{2} \le \alpha \le \frac{(n-j+1)}{2}\right\}, \\ & \text{if } j \in O[3k+1, n-1] \\ \{[2j-n+1, n-1]\} \cup \left\{[j-2k+1+2\alpha, 6k-j+2\alpha] : \right. \\ \qquad \left. \frac{(2k-j)}{2} \le \alpha \le \frac{(j-2k-2)}{2}\right\}, & \text{if } j \in E[2k+2, 3k] \\ \left\{[j-2k+1+2\alpha, 6k-j+2\alpha] : \frac{(j-n)}{2} \le \alpha \le \frac{(n-j)}{2}\right\}, \\ & \text{if } j \in E[3k+1, n-2]. \end{cases}$$

Then there exists $V \in T_j$ such that for all $p \in \mathbb{N} \cup \{0\}$

$$C^p_{Y_j} \le C^p_V.$$

PROOF. We will deal with the case where $j \in O[1, 2k-1]$. The other cases follow similarly. For $j \in O[1, k-1]$ we note by Lemma 5.10 that $V_j \subseteq [2k+2\beta - 2j, 2k+2\beta-1]$ for some β such that $0 \le \beta \le \min\{k, j, n-j\} = j$. If we set $\alpha = (2\beta - 1 - j)/2$, we note that α is an integer, $[2k-j+1+2\alpha, 2k+j+2\alpha] = [2k+2\beta-2j, 2k+2\beta-1]$ and $0 \le \beta \le j \implies (-1-j)/2 \le \alpha \le (j-1)/2$. Hence setting $V = [2k+2\beta-2j, 2k+2\beta-1]$ gives

$$C^p_{Y_j} \le C^p_{V_j} \text{ (by Lemma 5.10)} \le C^p_V \text{ (since } V_j \subseteq V).$$

(In fact $V_j = [2k+2\beta-2j, 2k+2\beta-1]$ in the above case so actually $V = V_j$.) On the other hand if $j \in O[k, 2k-1]$ we have $V_j = [2k+2\beta-2j, 2k+2\beta-1] \cap [1, n-1]$ for some β such that $0 \le \beta \le \min\{k, j, n-j\} = k$. Now if $0 \le \beta \le j-k$ we see $V_j \subseteq [1, 2j-1]$ and we set $V = [1, 2j-1]$. If $j-k+1 \le \beta \le k$ we see $V_j = [2k+2\beta-2j, 2k+2\beta-1]$ and if again we set $\alpha = (2\beta-1-j)/2$, α is an integer, $[2k-j+1+2\alpha, 2k+j+2\alpha] = V_j$ and $j-k+1 \le \beta \le k \implies (j+1-2k)/2 \le \alpha \le (2k-1-j)/2$ so we set $V = V_j$. Again since $V_j \subseteq V$ we have $C^p_{Y_j} \le C^p_{V_j} \le C^p_V$. □

For the record, if $j \in E[2, 2k-2]$ we set $V = V_j$ if $\beta \ge j-k+1$ and $V = [1, 2j-1]$ if $\beta \le j-k$ and set $\alpha = (2\beta - j)/2$. For $j \in [2k+1, n-1]$ we set $V = V_j$ if

$\beta \geq 3k - j + 1$ and $V = [2j - n + 1, n - 1]$ if $\beta \leq 3k - j$ while we set

$$\alpha = \begin{cases} \frac{n-2\beta+1-j}{2}, & \text{if } j \text{ is odd} \\ \frac{n-2\beta-j}{2}, & \text{if } j \text{ is even.} \end{cases}$$

It is useful to note here that V is always constructed from V_j in such a way that $V_j \subseteq V$. In fact, $V_j = V$ unless

(i) $j \in [k, 2k-1]$, $V = [1, 2j-1]$ and V_j has less than $2j - 1$ elements

or

(ii) $j \in [2k+1, 3k]$, $V = [2j - n + 1, n - 1]$ and V_j has less than $2(n-j) - 1$ elements.

We note that when $\alpha = 0$, the element V of T_j which arises is $S_j(A)$ which is also the V_j constructed from Y_j in Lemma 5.10, and in general, $V = S_j(A)$ if and only if $V_j = S_j(A)$. This may create a problem. Recall that we eventually hope to use an inductive hypothesis to establish results like $B^p_{X_j} \geq C^p_{Y_j}$ for $C \in \Gamma$. As Γ is still a large class of matrices, for any j there will be a large number of possibilities for Y_j. The number of possibilities for V_j is considerably smaller, whilst the number of elements in T_j is either the same or smaller. In light of this we will generally choose to show that either $B^p_{X_j} \geq C^p_{V_j}$ or $B^p_{X_j} \geq C^p_V$ rather than $B^p_{X_j} \geq C^p_{Y_j}$ even though in some cases this is a much tougher problem. If $V = V_j = S_j(A)$ showing that $B^p_{X_j} \geq C^p_V$ is potentially the most difficult problem. (In most cases $X_j = S_j(A)$, however the really difficult cases are when $j \in E[2, k-1] \cup O[3k+1, n-1]$). Of course if $Y_j = V_j = S_j(A)$ there is no way out, but if $Y_j \neq V_j = S_j(A) = V$ we may be making trouble for ourselves by "replacing" Y_j by V. We would still like to reduce the number of cases however, and the next lemma suggests an alternative "replacement" for Y_j to achieve this end.

LEMMA 5.12. *Let C, j, Γ and V_j be as in Lemma 5.10, and let $U_j = \{i \in [1, n-1] : a_{i\,j} = 1\} = S_j(A)$ (so if column $A^{(j)}$ is $\langle u, v\rangle$ then $U_j = [u, v]$). If $j \neq 2k$ and $Y_j \neq U_j$ but $V_j = U_j$, then for all $p \in \mathbb{N} \cup \{0\}$*

$$C^p_{Y_j} \leq C^p_{V'_j} \quad \text{or} \quad C^p_{Y_j} \leq C^p_{V''_j}$$

where

(a) *If $j \in E[2, 2k-2]$; that is, $Y_j \neq U_j = V_j = [2k - j, 2k + j - 1]$, then*

$V'_j = ([2k - j - 2, 2k - j - 1] \cup [2k - j + 2, 2k + j - 1]) \cap [1, n-1]$
(the intersection with $[1, n-1]$ is only significant for $j = 2k - 2$),
$V''_j = [2k - j, 2k + j - 3] \cup [2k + j, 2k + j + 1]$.

(b) *If $j \in O[1, 2k-1]$; that is, $Y_j \neq U_j = V_j = [2k - j + 1, 2k + j]$, then*

$V'_1 = V''_1 = [2k+2, 2k+3] \quad (U_1 = V_1 = [2k, 2k+1] \neq Y_1)$,
$V'_{2k-1} = V''_{2k-1} = \{1\} \cup [4, n-1] \quad (U_{2k-1} = V_{2k-1} = [2, n-1] \neq Y_{2k-1})$,
$V'_j = [2k - j - 1, 2k - j] \cup [2k - j + 3, 2k + j], \quad 1 < j < 2k - 1$,
$V''_j = [2k - j + 1, 2k + j - 2] \cup [2k + j + 1, 2k + j + 2], \quad 1 < j < 2k - 1$.

(c) *If* $j \in E[2k+2, n-2]$*; that is,* $Y_j \neq U_j = V_j = [j-2k+1, 6k-j]$*, then*

$$V_j' = [j-2k-1, j-2k] \cup [j-2k+3, 6k-j],$$
$$V_j'' = ([j-2k+1, 6k-j-2] \cup [6k-j+1, 6k-j+2]) \cap [1, n-1]$$

(the intersection with $[1, n-1]$ *is only significant for* $j = 2k+2$*).*

(d) *If* $j \in O[2k+1, n-1]$*; that is,* $Y_j \neq U_j = V_j = [j-2k, 6k-j-1]$*, then*

$$V_{n-1}' = V_{n-1}'' = [2k-3, 2k-2] \quad (U_{n-1} = V_{n-1} = [2k-1, 2k] \neq Y_{n-1}),$$
$$V_{2k+1}' = V_{2k+1}'' = [1, n-4] \cup \{n-1\} \quad (U_{2k+1} = V_{2k+1} = [1, n-2] \neq Y_{2k+1}),$$
$$V_j' = [j-2k-2, j-2k-1] \cup [j-2k+2, 6k-j-1], \quad 2k+1 < j < n-1,$$
$$V_j'' = [j-2k, 6k-j-3] \cup [6k-j, 6k-j+1], \quad 2k+1 < j < n-1.$$

PROOF. We will prove the case (b). Other cases may be proved by similar arguments. Note that by Lemma 5.10, since $[2k-j+1, 2k+j] = V_j = [2k+2\beta-2j, 2k+2\beta-1] \cap [1, n-1]$, then $2\beta = j+1$; that is, the number of elements in $Y_j \cap [2k, n-1]$ is $j+1$. Further, since there are at most $2j$ elements in Y_j there are at most $j-1$ elements in $Y_j \cap [1, 2k-1]$.

Suppose $j = 1$. We know there are two elements in $Y_1 \cap [2k, n-1]$ and $Y_1 \cap [1, 2k-1]$ is empty. Further $Y_1 \neq V_1 = [2k, 2k+1]$. If $2k+1 \in Y_1$, then $c_{2k+1\,1} = 1$ and so by condition 3 of Γ, $c_{2k\,1} = 1$ also. But this would mean $Y_1 = [2k, 2k+1]$, a contradiction. Hence $c_{2k+1\,1} = 0$ and $c_{2k\,1} = 0$. It follows that there are two elements in $Y_1 \cap [2k+2, n-1]$ and in fact that $Y_1 \subseteq [2k+2, n-1]$. It is now evident there is an injection $\rho : Y_1 \to [2k+2, 2k+3]$ such that $2k \leq \rho(i) \leq i$ for all $i \in Y_1$. Thus, by Corollary 5.7, $C^p_{Y_1} \leq C^p_{[2k+2,2k+3]} = C^p_{V_1'} = C^p_{V_1''}$ for all $p \in \mathbb{N} \cup \{0\}$.

Now suppose $j = 2k-1$. We know there are $j+1 = 2k$ elements in $Y_{2k-1} \cap [2k, n-1]$, so $[2k, n-1] \subseteq Y_{2k-1}$. Further, there are at most $j-1 = 2k-2$ elements in $Y_{2k-1} \cap [1, 2k-1]$. As there are $2k-1$ elements in $[1, 2k-1]$, there is at least one $i \in [1, 2k-1]$ such that $c_{i\,2k-1} = 0$. Note that by property 3 of Γ, if $i \in O[3, 2k-1]$ (respectively, $i \in E[2, 2k-2]$) then also $c_{i-1\,2k-1} = 0$ (respectively, $c_{i+1\,2k-1} = 0$) and $i-1 \in E[2, 2k-2]$ (respectively, $i+1 \in O[3, 2k-1]$). Thus for there to be exactly one i such that $c_{i\,2k-1} = 0$, $i = 1$ and so $Y_{2k-1} = [2, n-1] = V_{2k-1}$ which is a contradiction. It follows that there is at least one $i' \in O[3, 2k-1]$ such that $c_{i'-1\,2k-1} = c_{i'\,2k-1} = 0$; that is, $Y_{2k-1} \cap [1, 2k-1] \subseteq [1, i'-2] \cup [i'+1, 2k-1]$ for some $i' \in O[3, 2k-1]$. Thus there is an injection $\rho_1 : Y_{2k-1} \cap [1, 2k-1] \to \{1\} \cup [4, 2k-1] = V_{2k-1}' \cap [1, 2k-1]$ such that $i'' \leq \rho_1(i'') \leq 2k-1$ for all $i'' \in Y_{2k-1} \cap [1, 2k-1]$.

Since $k \geq 2$ (recall the case $k = 1$ requires no proof since the permutation of maximum entropy is a cycle when $k = 1$), $V_{2k-1}' \cap [2k, n-1] = [2k, n-1] = Y_{2k-1} \cap [2k, n-1]$. Thus there is an injection $\rho : Y_{2k-1} \to V_{2k-1}' = V_{2k-1}''$ such that $i'' \leq \rho(i'') \leq 2k-1$ or $2k \leq \rho(i'') = i''$ for all $i'' \in Y_{2k-1}$. Thus, by Corollary 5.7

$$C^p_{Y_{2k-1}} \leq C^p_{V_{2k-1}'} \leq C^p_{V_{2k-1}''}$$

for all $p \in \mathbb{N} \cup \{0\}$.

Finally, suppose $1 < j < 2k-1$. We know we have $j+1$ elements in $Y_j \cap [2k, n-1]$ and at most $j-1$ elements in $Y_j \cap [1, 2k-1]$. In the instance where $Y_j \cap [2k, n-1] = [2k, 2k+j] = V_j \cap [2k, n-1] = V_j' \cap [2k, n-1]$ note that $Y_j \cap [2k-j+1, 2k-1] \neq [2k-j+1, 2k-1]$ since otherwise we would have $Y_j = [2k-j+1, 2k+1] = V_j$. Thus there is at least one $i \in [2k-j+1, 2k-1]$ such that $c_{i\,j} = 0$. Using property 3 of Γ and the fact that $2k-j+1$ is even and a similar argument to the above establishes the existence of at least one $i' \in O[2k-j+2, 2k-1]$ such that $Y_j \cap [2k-j+1, 2k-1] \subseteq [2k-j+1, i'-2] \cup [i'+1, 2k-1]$ (note that if $i' = 2k-j+2$, $[2k-j+1, i'-2] = \emptyset$ and if $i' = 2k-1$, $[i'+1, 2k-1] = \emptyset$). Now $V_j' \cap [2k-j+1, 2k-1] = [2k-j+3, 2k-1]$, which contains the $j-3$ elements "most centrally spaced"; that is, closest to $2k-1$ in $[2k-j+1, 2k-1]$, while $Y_j \cap [2k-j+1, 2k-1]$ contains at most $j-3$ elements. Also, $V_j' \cap [1, 2k-j] = [2k-j-1, 2k-j]$ which contains the two elements "most centrally spaced" in $[1, 2k-j]$. Combining this with the fact that $Y_j \cap [1, 2k-1]$ has at most $j-1$ elements, we deduce the existence of an injection $\rho : Y_j \cap [1, 2k-1] \to V_j' \cap [1, 2k-1]$ such that $i'' \leq \rho(i'') \leq 2k-1$ for all $i'' \in Y_j \cap [1, 2k-1]$. As $Y_j \cap [2k, n-1] = V_j' \cap [2k, n-1]$ this is enough to apply Corollary 5.7 and deduce $C^p_{Y_j} \leq C^p_{V_j'}$ for all $p \in \mathbb{N} \cup \{0\}$.

In the instance where $Y_j \cap [2k, n-1] \neq [2k, 2k+j]$ we use the fact that $Y_j \cap [2k, n-1]$ has $j+1$ elements and exploit property 3 of Γ in the now familiar manner to establish first the existence of at least one $i' \in O[2k+1, 2k+j]$ such that $Y_j \cap [2k, 2k+j] \subseteq [2k, i'-2] \cup [i'+1, 2k+j]$, and then the fact that $V_j'' \cap [2k, 2k+j] = [2k, 2k+j-2]$ and $V_j'' \cap [2k+j+1, n-1] = [2k+j+1, 2k+j+2]$ to deduce the existence of an injection $\rho_2 : Y_j \cap [2k, n-1] \to V_j'' \cap [2k, n-1]$ such that $2k \leq \rho_2(i'') \leq i''$ for all $i'' \in Y_j \cap [2k, n-1]$. Now as $Y_j \cap [1, 2k-1]$ has at most $j-1$ elements and $V_j'' \cap [1, 2k-1] = [2k-j+1, 2k-1]$ (the "most central" $j-1$ element subset of $[1, 2k-1]$), we see we also have an injection $\rho_1 : Y_j \cap [1, 2k-1] \to V_j'' \cap [1, 2k-1]$ such that $i'' \leq \rho_1(i'') \leq 2k-1$ for all $i'' \in Y_j \cap [1, 2k-1]$. Combining the injections and applying Corollary 5.7 gives $C^p_{Y_j} \leq C^p_{V_j''}$ for all $p \in \mathbb{N} \cup \{0\}$ as required. □

Observe that we have shown that if $j \in [1, 2k-1] \cup [2k+1, n-1]$ and for each $C \in \Gamma$ if $Y_j = S_j(C)$ then there exists $V \in T_j$ such that for all $p \in \mathbb{N} \cup \{0\}$

$$C^p_{Y_j} \leq C^p_V \quad \text{(Corollary 5.11)}.$$

Further to this, if $Y_j \neq U_j = S_j(A)$ but $V_j = U_j$ where V_j is constructed from Y_j as in Lemma 5.10, then for all $p \in \mathbb{N} \cup \{0\}$

$$C^p_{Y_j} \leq C^p_{V_j'} \quad \text{or} \quad C^p_{Y_j} \leq C^p_{V_j''},$$

where V_j' and V_j'' are defined in Lemma 5.12. These observations may be combined to form the following main corollary used in the proof of the main result.

COROLLARY 5.13. *Let $C \in \Gamma$ and let $j \in [1, 2k-1] \cup [2k+1, n-1]$, $Y_j = S_j(C)$, $U_j = S_j(A)$, T_j be as defined in Corollary 5.11 and V_j' and V_j'' be defined as in Lemma 5.12. Define T_j' by*

$$T_j' = (T_j \setminus \{U_j\}) \cup \{V_j', V_j{}''\}.$$

Then there exists $W \in T_j$ such that $C^p_{Y_j} \leq C^p_W$ for all $p \in \mathbb{N} \cup \{0\}$. Further, if $Y_j \neq U_j$ then there exists $W \in T_j'$ such that $C^p_{Y_j} \leq C^p_W$ for all $p \in \mathbb{N} \cup \{0\}$.

PROOF. Let V_j and V be the sets constructed from Y_j in Lemma 5.10 and Corollary 5.11 respectively and recall our remark following Corollary 5.11 that $V = U_j$ if and only if $V_j = U_j$. By Corollary 5.11 if we wish to find $W \in T_j$ such that $C^p_{Y_j} \leq C^p_W$ for all $p \in \mathbb{N} \cup \{0\}$ we may choose W to be V. Further, if $V \neq U_j$ then to find a $W \in T'_j$ such that $C^p_{Y_j} \leq C^p_W$ for all $p \in \mathbb{N} \cup \{0\}$ we may still choose W to be V. Finally, if $V = U_j$, and so consequently $V_j = U_j$ but $Y_j \neq U_j$, an appeal to Lemma 5.12 shows that there is at least one choice of V'_j or V''_j for W; that is, there is a $W \in T'_j$ such that $C^p_{Y_j} \leq C^p_W$ for all $p \in \mathbb{N} \cup \{0\}$. □

The notation used in the above result and in its supporting results Lemmas 5.10, 5.12 and Corollary 5.11 has proved useful in efficiently specifying these results. However as mentioned earlier, we will use an alternative notation for the proofs of our main results. It is clear that the distribution of 1's in a given column of a matrix C has great bearing on the sum $C^p_{Y_j}$. Furthermore, the proximity of these 1's to $2k$ is of vital importance. Thus the notation we will use now describes the column with the emphasis on the central element $2k$. For example, for $j \in E[3k+1, n-2]$ we have $B^{(j)} = \langle j - 2k + 1, 6k - j \rangle$ in our previous notation and $B^{(j)} = B^{(n-2i)} = \langle 2k - 2i + 1, 2k + 2i \rangle$ for $1 \leq i \leq (k-1)/2$ in the new notation. We now see immediately that the unit elements of $B^{(j)}$ are as close to $2k$ as possible. This fact is not so obvious in the previous notation. With this in mind we respecify T_j and T'_j broken up as they are actually used in our main inductive proofs. Note that we do not need to specify T_j for $j \in E[2, k-1]$ or $j \in O[3k+1, n-1]$ as we need to treat the cases $Y_j = U_j$ quite separately here than elsewhere. (These being the major problem cases in which $X_j \neq U_j$.)

To assist the reader in the cases where the old notation may make the statement of the desired results easier to understand than the new notation used in the proof of the results, we will state the results as we prove them (that is, with the new notation first, and then immediately restate the results in the old notation).

REPRESENTATION 5.14. The following new representation is given for the collection of subsets T_j and T'_j defined earlier in Corollaries 5.11 and 5.13.

(i)

$$T_1 = \{[2k-2, 2k-1], [2k, 2k+1]\}$$

$$\begin{aligned} T_{2i-1} = &\{[2k - 2i + 2 + 2\alpha, 2k + 2i - 1 + 2\alpha] : \alpha \in [1, i-1]\} \\ &\cup \{[2k - 2i + 2 - 2\alpha, 2k + 2i - 1 - 2\alpha] : \alpha \in [1, i]\} \\ &\cup \{[2k - 2i + 2, 2k + 2i - 1]\}, \quad \text{for } i \in [2, k/2]. \end{aligned}$$

$$T'_1 = \{[2k-2, 2k-1], [2k+2, 2k+3]\}$$

$$\begin{aligned} T'_{2i-1} = &(T_{2i-1} \setminus \{[2k - 2i + 2, 2k + 2i - 1]\}) \\ &\cup \{[2k - 2i, 2k - 2i + 1] \cup [2k - 2i + 4, 2k + 2i - 1]\} \\ &\cup \{[2k - 2i + 2, 2k + 2i - 3] \cup [2k + 2i, 2k + 2i + 1]\}, \\ &\text{for } i \in [2, k/2]. \end{aligned}$$

(ii)

$$\begin{aligned} T_{n-2i} = &\{[2k-2i+1+2\alpha, 2k+2i+2\alpha] : \alpha \in [1,i]\} \\ &\cup \{[2k-2i+1-2\alpha, 2k+2i-2\alpha] : \alpha \in [1,i]\} \\ &\cup \{[2k-2i+1, 2k+2i]\}, \quad \text{for } i \in [1,(k-1)/2]. \end{aligned}$$

$$\begin{aligned} T'_{n-2i} = &(T_{n-2i} \setminus \{[2k-2i+1, 2k+2i]\}) \\ &\cup \{[2k-2i-1, 2k-2i] \cup [2k-2i+3, 2k+2i]\} \\ &\cup \{[2k-2i+1, 2k+2i-2] \cup [2k+2i+1, 2k+2i+2]\}, \\ &\text{for } i \in [1,(k-1)/2]. \end{aligned}$$

(iii)

$$\begin{aligned} T'_{2i} = &\{[2k-2i+2\alpha, 2k+2i-1+2\alpha] : \alpha \in [1,i]\} \\ &\cup \{[2k-2i-2\alpha, 2k+2i-1-2\alpha] : \alpha \in [1,i]\} \\ &\cup \{[2k-2i-2, 2k-2i-1] \cup [2k-2i+2, 2k+2i-1]\} \\ &\cup \{[2k-2i, 2k+2i-3] \cup [2k+2i, 2k+2i+1]\}, \\ &\text{for } i \in [1,(k-1)/2]. \end{aligned}$$

(iv)

$$T'_{n-1} = \{[2k+1, 2k+2], [2k-3, 2k-2]\}$$

$$\begin{aligned} T'_{n-2i+1} = &\{[2k-2i+1+2\alpha, 2k+2i-2+2\alpha] : \alpha \in [1,i]\} \\ &\cup \{[2k-2i+1-2\alpha, 2k+2i-2-2\alpha] : \alpha \in [1,i-1]\} \\ &\cup \{[2k-2i-1, 2k-2i] \cup [2k-2i+3, 2k+2i-2]\} \\ &\cup \{[2k-2i+1, 2k+2i-4] \cup [2k+2i-1, 2k+2i]\}, \\ &\text{for } i \in [2,k/2]. \end{aligned}$$

(v)

$$T_{2k+1} = \{[3, n-1], [1, n-2]\}$$

$$\begin{aligned} T_{2k+2i-1} = &\{[2i-1+2\alpha, n-2i+2\alpha] : \alpha \in [1,i-1]\} \\ &\cup \{[2i-1-2\alpha, n-2i-2\alpha] : \alpha \in [1,i-1]\} \\ &\cup \{[4i-1, n-1]\} \cup \{[2i-1, n-2i]\}, \quad \text{for } i \in [2,(k+1)/2]. \end{aligned}$$

$$T'_{2k+1} = \{[3, n-1], [1, n-4] \cup \{n-1\}\}$$

$$\begin{aligned} T'_{2k+2i-1} = &(T_{2k+2i-1} \setminus \{[2i-1, n-2i]\}) \\ &\cup \{[2i-3, 2i-2] \cup [2i+1, n-2i]\} \\ &\cup \{[2i-1, n-2i-2] \cup [n-2i+1, n-2i+2]\}, \end{aligned}$$

for $i \in [2, (k+1)/2]$.

(vi)

$$T_{2k+2} = \{[1, n-4], [5, n-1], [3, n-2]\}$$

$$\begin{aligned} T_{2k+2i} = &\{[2i+1+2\alpha, n-2i+2\alpha] : \alpha \in [1, i-1]\} \\ &\cup \{[2i+1-2\alpha, n-2i-2\alpha] : \alpha \in [1, i]\} \\ &\cup \{[4i+1, n-1]\} \cup \{[2i+1, n-2i]\}, \\ &\text{for } i \in [2, k/2]. \end{aligned}$$

$$\begin{aligned} T'_{2k+2i} = &(T_{2k+2i} \setminus \{[2i+1, n-2i]\}) \cup \{[2i-1, 2i] \cup [2i+3, n-2i]\} \\ &\cup \{[1, n-1] \cap ([2i+1, n-2i-2] \cup [n-2i+1, n-2i+2])\}, \\ &\text{for } i \in [1, k/2]. \end{aligned}$$

(vii)

$$T_{2k-1} = \{[1, n-3], [2, n-1]\}$$

$$\begin{aligned} T_{2k-2i+1} = &\{[2i+2\alpha, n-2i+1+2\alpha] : \alpha \in [1, i-1]\} \\ &\cup \{[2i-2\alpha, n-2i+1-2\alpha] : \alpha \in [1, i-1]\} \\ &\cup \{[1, n-4i+1]\} \cup \{[2i, n-2i+1]\}, \\ &\text{for } i \in [2, (k+1)/2]. \end{aligned}$$

$$T'_{2k-1} = \{[1, n-3], \{1\} \cup [4, n-1]\}$$

$$\begin{aligned} T'_{2k-2i+1} = &(T_{2k-2i+1} \setminus \{[2i, n-2i+1]\}) \\ &\cup \{[2i-2, 2i-1] \cup [2i+2, n-2i+1]\} \\ &\cup \{[2i, n-2i-1] \cup [n-2i+2, n-2i+3]\}, \\ &\text{for } i \in [2, (k+1)/2]. \end{aligned}$$

(viii)

$$T_{2k-2} = \{[4, n-1], [1, n-5], [2, n-3]\}$$

$$\begin{aligned} T_{2k-2i} = &\{[2i+2\alpha, n-2i-1+2\alpha] : \alpha \in [1, i]\} \\ &\cup \{[2i-2\alpha, n-2i-1-2\alpha] : \alpha \in [1, i-1]\} \\ &\cup \{[1, n-4i-1]\} \cup \{[2i, n-2i-1]\}, \\ &\text{for } i \in [2, k/2]. \end{aligned}$$

$$T'_{2k-2i} = (T_{2k-2i} \setminus \{[2i, n-2i-1]\})$$

$$\cup \{([2i-2, 2i-1] \cup [2i+2, n-2i-1]) \cap [1, n-1]\}$$
$$\cup \{[2i, n-2i-3] \cup [n-2i, n-2i+1]\},$$
$$\text{for } i \in [1, k/2].$$

6. Two inductive lemmas

In this section we show that the norm $\|B^p\|$ dominates the norm $\|C^p\|$, where $C \in \Gamma$. The arguments we use here are inductive. Because of their repetitive nature, we have devised three algorithms to do the job in a more concise way.

NOTATION 6.1. We introduce the following notation for Lemmas 45 and 47. For any matrices C and D,

$$C^m_{[a,b]} = \sum_{i=a}^{b} \left|(C^m)^{(i)}\right|$$

and

$$(C^m - D^m)[a, b] = \sum_{i=a}^{b} \left|(C^m)^{(i)}\right| - \sum_{i=a}^{b} \left|(D^m)^{(i)}\right|.$$

LEMMA 6.2. *Let $B = M(\theta_n)$ and $m \in \mathbb{N}$, then*

(i) $\left|(B^m)^{(2i-1)}\right| \geq \left|(B^m)^{(n-2i+1)}\right|$ *for all* $i \in [1, k/2]$

(ii) $\left|(B^m)^{(k)}\right| = \left|(B^m)^{(3k)}\right|$

(iii) $\left|(B^m)^{(2k+2i-1)}\right| \geq \left|(B^m)^{(2k-2i+1)}\right|$ *for all* $i \in [1, k/2]$

(iv) $\left|(B^m)^{(n-2i)}\right| \geq \left|(B^m)^{(2i)}\right|$ *for all* $i \in [1, (k-1)/2]$

(v) $\left|(B^m)^{(2k+2i)}\right| \geq \left|(B^m)^{(2k-2i)}\right|$ *for all* $i \in [1, (k-1)/2]$

(vi) $\left|(B^m)^{(2k-2i+2)}\right| \geq \left|(B^m)^{(2k+2i-1)}\right|$ *for all* $i \in [1, (k+1)/2]$

(vii) $\left|(B^m)^{(n-2i-1)}\right| \geq \left|(B^m)^{(n-2i)}\right|$ *for all* $i \in [1, (k-1)/2]$

(viii) $\left|(B^m)^{(2k-2i+1)}\right| \geq \left|(B^m)^{(2k+2i)}\right|$ *for all* $i \in [1, k/2]$

(ix) $\left|(B^m)^{(2i)}\right| \geq \left|(B^m)^{(2i-1)}\right|$ *for all* $i \in [1, k/2]$.

LEMMA 6.3. *Restatement of Lemma 6.2 and Corollary. Let $B = M(\theta_n)$ and $m \in \mathbb{N}$, then*

1. $\left|(B^m)^{(k)}\right| = \left|(B^m)^{(3k)}\right|$
2. *For* $j \in O[1, k-1]$, $\left|(B^m)^{(n-j)}\right| \leq \left|(B^m)^{(j)}\right| \leq \left|(B^m)^{(j+1)}\right|$
3. *For* $j \in E[2, k-1]$, $\left|(B^m)^{(j)}\right| \leq \left|(B^m)^{(n-j)}\right| \leq \left|(B^m)^{(n-j-1)}\right|$
4. *For* $j \in [k+1, 2k]$, $\left|(B^m)^{(n-j+1)}\right| \leq \left|(B^m)^{(j)}\right| \leq \left|(B^m)^{(n-j)}\right|$
5. *For* $j \in [1, k-1]$, $\left|(B^m)^{(j)}\right| \leq \left|(B^m)^{(n-j-1)}\right|$ *and* $\left|(B^m)^{(n-j)}\right| \leq \left|(B^m)^{(j+1)}\right|$.

(1 is 6.2 (ii), 2 is 6.2 (i) and (ix), 3 is 6.2 (iv) and (vii) and 4 is 6.2 (iii),(v), (vi) and (viii) while 5 is a corollary of 2 and 3.)

Lemmas 6.2 and 6.3 replace the use of cones in earlier results of our main type. Note that if we abbreviate $\left|(B^m)^{(j)}\right|$ to $[j]$ we can summarise a complete ranking of the jth column sums of B^m in decreasing order as follows:

$$[2k] \geq [2k+1] \geq [2k-1] \geq [2k+2] \geq [2k-2] \geq \dots$$
$$\dots \geq [3k+2] \geq [k-2] \geq [3k+1] \geq [k-1] \geq [k] = [3k]$$

and

$$[k] = [3k] \geq [3k+1] \geq [k-1] \geq [k-2] \geq [3k+2] \geq \dots$$
$$\dots \geq [n-4] \geq [4] \geq [3] \geq [n-3] \geq [n-2] \geq [2] \geq [1] \geq [n-1]$$

for k odd, and

$$[3k] = [k] \geq [k-1] \geq [3k+1] \geq [3k+2] \geq [k-2] \geq \dots$$
$$\dots \geq [n-4] \geq [4] \geq [3] \geq [n-3] \geq [n-2] \geq [2] \geq [1] \geq [n-1]$$

for k even.

PROOF OF LEMMA 6.2. Since $B \in \Gamma$, (ix) and (vii) follow by Corollary 5.6 and (vi) and (viii) follow by Lemma 5.5. The proof of the remainder is by induction. For any $p \geq 1$ let Claim (1, p), Claim (2, p), ..., Claim (5, p) be as follows:

Claim (1, p) $\left|(B^p)^{(2i-1)}\right| \geq \left|(B^p)^{(n-2i+1)}\right|$ for all $i \in [1, k/2]$.

Claim (2, p) $\left|(B^p)^{(k)}\right| = \left|(B^p)^{(3k)}\right|$.

Claim (3, p) $\left|(B^p)^{(2k+2i-1)}\right| \geq \left|(B^p)^{(2k-2i+1)}\right|$ for all $i \in [1, k/2]$.

Claim (4, p) $\left|(B^p)^{(n-2i)}\right| \geq \left|(B^p)^{(2i)}\right|$ for all $i \in [1, (k-1)/2]$.

Claim (5, p) $\left|(B^p)^{(2k+2i)}\right| \geq \left|(B^p)^{(2k-2i)}\right|$ for all $i \in [1, (k-1)/2]$.

For $p = 1$, $\left|B^{(j)}\right| = \left|B^{(n-j)}\right|$ for all $j \in \{1, \dots, n-1\}$ hence all claims are true.

We now assume all claims are true for $p = m$ (if $r \in [1, 5]$ and $s \in [1, m]$ we refer to the assumed true Claim (r, s) as inductive hypothesis (r, s)) and aim to show all claims are true for $p = m + 1$.

Proof of Claim (1, m+1). Let $i \in [1, k/2]$. Then

$$\begin{aligned}
&\left|\left(B^{m+1}\right)^{(2i-1)}\right| - \left|\left(B^{m+1}\right)^{(n-2i+1)}\right| \\
&\quad = B^m_{[2k-2i+2, 2k+2i-1]} - B^m_{[2k-2i+3, 2k+2i]} \\
&\quad = \left|(B^m)^{(2k-2i+2)}\right| - \left|(B^m)^{(2k+2i)}\right| \\
&\quad \geq 0
\end{aligned}$$

by (vi) and Corollary 5.6.

Proof of Claim (2, m+1).

$$\begin{aligned}
&\left|\left(B^{m+1}\right)^{(k)}\right| - \left|\left(B^{m+1}\right)^{(3k)}\right| \\
&\quad = \begin{cases} B^m_{[k+1,3k]} - B^m_{[k,3k-1]}, & \text{if } k \text{ is odd} \\ B^m_{[k,3k-1]} - B^m_{[k+1,3k]}, & \text{if } k \text{ is even} \end{cases} \\
&\quad = \begin{cases} \left|(B^m)^{(3k)}\right| - \left|(B^m)^{(k)}\right|, & \text{if } k \text{ is odd} \\ \left|(B^m)^{(k)}\right| - \left|(B^m)^{(3k)}\right|, & \text{if } k \text{ is even} \end{cases} \\
&\quad \geq 0
\end{aligned}$$

by inductive hypothesis (2, m).

Proof of Claim (3, m+1). Let $i \in [1, k/2]$. Then

$$\begin{aligned}
&\left|\left(B^{m+1}\right)^{(2k+2i-1)}\right| - \left|\left(B^{m+1}\right)^{(2k-2i+1)}\right| \\
&\quad = B^m_{[2i-1,n-2i]} - B^m_{[2i,n-2i+1]} \\
&\quad = \left|(B^m)^{(2i-1)}\right| - \left|(B^m)^{(n-2i+1)}\right| \\
&\quad \geq 0
\end{aligned}$$

by inductive hypothesis (1, m).

Proof of Claim (4, m+1). Let $i \in [1, (k-1)/2]$. Then

$$\begin{aligned}
&\left|\left(B^{m+1}\right)^{(n-2i)}\right| - \left|\left(B^{m+1}\right)^{(2i)}\right| \\
&\quad = B^m_{[2k-2i+1,2k+2i]} - B^m_{[2k-2i+2,2k+2i+1]} \\
&\quad = \left|(B^m)^{(2k-2i+1)}\right| - \left|(B^m)^{(2k+2i+1)}\right| \\
&\quad \geq 0
\end{aligned}$$

by (viii) and Corollary 5.6.

Proof of Claim (5, m+1). Let $i \in [1, (k-1)/2]$. Then

$$\begin{aligned}
&\left|\left(B^{m+1}\right)^{(2k+2i)}\right| - \left|\left(B^{m+1}\right)^{(2k-2i)}\right| \\
&\quad = B^m_{[2i+1,n-2i]} - B^m_{[2i,n-2i-1]} \\
&\quad = \left|(B^m)^{(n-2i)}\right| + \left|(B^m)^{(2i)}\right| \\
&\quad \geq 0
\end{aligned}$$

by inductive hypothesis (4, m). □

The next lemma is the key to the proof of the main theorem. The inductive statement is very complex so we give a brief overview of the aim of the various parts of this lemma.

Recall that our goal is to show that $\|B^p\| \geq \|C^p\|$ for all $p \in \mathbb{N}$ and for all $C \in \Gamma$. We have already stated that it is not true that $\left|(B^p)^{(j)}\right| \geq \left|(C^p)^{(j)}\right|$ for all

$j \in [1, n-1]$, but that it is true for most $j \in [1, n-1]$: indeed condition (v) says that the statement is true for all $j \in O[1, k-1] \cup \{k\} \cup [2k, 3k] \cup E[3k+1, n-2]$. Furthermore, the statement is also true for $j \in E[2, k-1] \cup O[3k+1, n-1]$ provided $C^{(j)} \neq A^{(j)}$ (conditions (v) (a) and (b)) and for $j \in [k+1, 2k-1]$ provided $C^{(j)} \neq B^{(j)}$ $\left(= A^{(j)}\right)$ or $C^{(n-j)} \neq B^{(n-j)}$ $\left(= A^{(n-j)}\right)$ (conditions (v) (c) and (d)). The cases excluded in conditions (v) (c) and (d) are dealt with in condition (vi) whilst those excluded in conditions (v) (a) and (b) are dealt with in condition (vii). In both of these cases, the problem is resolved by comparing the sums $\left|(B^m)^{(j)}\right| + \left|(B^m)^{(s)}\right|$ and $\left|(C^m)^{(j)}\right| + \left|(C^m)^{(s)}\right|$ for particular columns s of the matrices B and C.

Conditions (i) to (iv) are necessary to establish conditions (v) to (vii) and we combine them with Lemma 6.2 and Corollary 5.6 to form the concise result $\left|(B^m)^{(s)}\right| \geq \left|(C^m)^{(s')}\right|$ where $s \in [2, n-2]$ and $s' \in \{s-1, n-s+1\}$ for $s \leq 2k$ or $s' \in \{s+1, n-s-1\}$ for $s \geq 2k$.

The various column sum comparisons described in conditions (viii) to (xi) are again used as supporting results for conditions (v) to (vii). Of course, a corollary of condition (xi) is precisely the result we require.

Lemma 6.4. *Let $C \in \Gamma$ and for any $j \in [1, n-1]$, let $C^{(j)}$ be denoted by $\langle a, b\rangle^j$. (Note that this does not mean that $\langle C^{(j)}\rangle = [a, b]$.) Then for all $m \in \mathbb{N}$,*

(i) $\left|(B^m)^{(2i-1)}\right| \geq \left|(C^m)^{(n-2i+1)}\right|$ *for all $i \in [1, k/2]$*

(ii) $\left|(B^m)^{(n-2i)}\right| \geq \left|(C^m)^{(2i)}\right|$ *for all $i \in [1, (k-1)/2]$*

(iii) $\left|(B^m)^{(2k+2i-1)}\right| \geq \left|(C^m)^{(2k-2i+1)}\right|$ *for all $i \in [1, (k+1)/2]$*

(iv) $\left|(B^m)^{(2k+2i)}\right| \geq \left|(C^m)^{(2k-2i)}\right|$ *for all $i \in [1, k/2]$*

(v) $\left|(B^m)^{(j)}\right| \geq \left|(C^m)^{(j)}\right|$ *for all $j \in O[1, k-1] \cup \{k\} \cup [2k, 3k] \cup E[3k+1, n-2]$ and for all j such that*

(a) *$j = 2i$, with $i \in [1, (k-1)/2]$ and $\langle a, b\rangle^{2i} \neq \langle 2k-2i, 2k+2i-1\rangle$*

or

(b) *$j = n-2i+1$, with $i \in [1, k/2]$ and $\langle a, b\rangle^{n-2i+1} \neq \langle 2k-2i+1, 2k+2i-2\rangle$*

or

(c) *$j = 2k-2i+1$, with $i \in [1, k/2]$ and $\langle a, b\rangle^{2k-2i+1} \neq \langle 2i, n-2i+1\rangle$ or $\langle a, b\rangle^{2k+2i-1} \neq \langle 2i-1, n-2i\rangle$*

or

(d) *$j = 2k-2i$, with $i \in [1, (k-1)/2]$ and $\langle a, b\rangle^{2k-2i} \neq \langle 2i, n-2i-1\rangle$ or $\langle a, b\rangle^{2k+2i} \neq \langle 2i+1, n-2i\rangle$*

(vi) $\left|(B^m)^{(j)}\right| + \left|(B^m)^{(n-j)}\right| \geq \left|(C^m)^{(j)}\right| + \left|(C^m)^{(n-j)}\right|$ *for j such that*

(a) *$j = 2k-2i+1$, with $i \in [1, (k+1)/2]$ and $\langle a, b\rangle^{2k-2i+1} = \langle 2i, n-2i+1\rangle$ and $\langle a, b\rangle^{n-j} = \langle a, b\rangle^{2k+2i-1} = \langle 2i-1, n-2i\rangle$*

or

(b) *$j = 2k-2i$, with $i \in [1, k/2]$ and $\langle a, b\rangle^{2k-2i} = \langle 2i, n-2i-1\rangle$ and $\langle a, b\rangle^{n-j} = \langle a, b\rangle^{2k+2i} = \langle 2i+1, n-2i\rangle$*

(vii) $\left|(B^m)^{(j)}\right| + \left|(B^m)^{(s)}\right| \geq \left|(C^m)^{(j)}\right| + \left|(C^m)^{(s)}\right|$ *for $j \in E[2, k-1] \cup O[3k+1, n-1]$ and $s \in \{n-j, 2k-j, 2k+j : j \in E[2, k-1]\}$ or $s \in \{n-j, j-2k, 6k-j : j \in O[3k+1, n-1]\}$ such that $C^{(j)}$ and $C^{(s)}$ satisfy one of the following conditions:*

(a) $\langle a,b\rangle^{2i} = \langle 2k-2i, 2k+2i-1\rangle$ *and* $\langle a,b\rangle^{n-2i} \neq \langle 2k-2i+1, 2k+2i\rangle$ *for* $i \in [1,(k-1)/2]$

(b) $\langle a,b\rangle^{2i} = \langle 2k-2i, 2k+2i-1\rangle$ *and* $\langle a,b\rangle^{2k-2i} \neq \langle 2i, n-2i-1\rangle$ *for* $i \in [1,(k-1)/2]$

(c) $\langle a,b\rangle^{2i} = \langle 2k-2i, 2k+2i-1\rangle$ *and* $\langle a,b\rangle^{2k+2i} \neq \langle 2i+1, n-2i\rangle$ *for* $i \in [1,(k-1)/2]$

(d) $\langle a,b\rangle^{n-2i+1} = \langle 2k-2i+1, 2k+2i-2\rangle$ *and* $\langle a,b\rangle^{2i-1} \neq \langle 2k-2i+2, 2k+2i-1\rangle$ *for* $i \in [1,k/2]$

(e) $\langle a,b\rangle^{n-2i+1} = \langle 2k-2i+1, 2k+2i-2\rangle$ *and* $\langle a,b\rangle^{2k-2i+1} \neq \langle 2i, n-2i+1\rangle$ *for* $i \in [1,k/2]$

(f) $\langle a,b\rangle^{n-2i+1} = \langle 2k-2i+1, 2k+2i-2\rangle$ *and* $\langle a,b\rangle^{2k+2i-1} \neq \langle 2i-1, n-2i\rangle$ *for* $i \in [1,k/2]$

(viii) $B^m_{[j,n-j]} \geq C^m_{[j,n-j]}$ *for all* $j \in [k,2k]$

(ix) $\left|(B^m)^{(j)}\right| + \left|(B^m)^{(2k-j)}\right| + \left|(B^m)^{(2k+j)}\right| + \left|(B^m)^{(n-j)}\right| \geq \left|(C^m)^{(j)}\right| + \left|(C^m)^{(2k-j)}\right| + \left|(C^m)^{(2k+j)}\right| + \left|(C^m)^{(n-j)}\right|$ *for all* $j \in [1,k-1]$

(x) $B^m_{[j,2k-j]\cup[2k+j,n-j]} \geq C^m_{[j,2k-j]\cup[2k+j,n-j]}$ *for all* $j \in [1,k-1]$

(xi) $B^m_{[j,n-j]} \geq C^m_{[j,n-j]}$ *for all* $j \in [1,2k]$.

Restatement of Lemma 6.4.

Let $C \in \Gamma$ and for any $j \in [1,n-1]$ let $Y_j = S_j(C)$, $U_j = S_j(A)$ and $X_j = S_j(B)$. Then for all $m \in \mathbb{N}$

1. $\left|(B^m)^{(j)}\right| \geq \left|(C^m)^{(n-j)}\right|$ for all $j \in O[1,k-1]\cup E[3k+1,n-2]\cup[2k+1,3k]$.
2. If $j \in [1,n-1]$ then $\left|(B^m)^{(j)}\right| \geq \left|(C^m)^{(j)}\right|$ unless

(2.1) $j \in E[2,k-1] \cup O[3k+1,n-1]$ and $Y_j = U_j$, or unless

(2.2) $j \in [k+1,2k-1]$ and $Y_j = U_j = X_j$ and $Y_{n-j} = U_{n-j} = X_{n-j}$.

3. $\left|(B^m)^{(j)}\right| + \left|(B^m)^{(n-j)}\right| \geq \left|(C^m)^{(j)}\right| + \left|(C^m)^{(n-j)}\right|$ for all $j \in [k+1,2k-1]$ such that $Y_j = U_j = X_j$ and $Y_{n-j} = U_{n-j} = X_{n-j}$.
4. $\left|(B^m)^{(j)}\right| + \left|(B^m)^{(s)}\right| \geq \left|(C^m)^{(j)}\right| + \left|(C^m)^{(s)}\right|$ for

(4.1) $j \in E[2,k-1]$ and $s \in \{n-j, 2k-j, 2k+j\}$ where $Y_j = U_j$ but $Y_s \neq U_s = X_s$, and for

(4.2) $j \in O[3k+1,n-1]$ and $s \in \{n-j, j-2k, 6k-j\}$ where $Y_j = U_j$ but $Y_s \neq U_s = X_s$.

5. $B^m_{[j,n-j]} \geq C^m_{[j,n-j]}$ for all $j \in [k,2k]$.
6. $\left|(B^m)^{(j)}\right| + \left|(B^m)^{(2k-j)}\right| + \left|(B^m)^{(2k+j)}\right| + \left|(B^m)^{(n-j)}\right| \geq \left|(C^m)^{(j)}\right| + \left|(C^m)^{(2k-j)}\right| + \left|(C^m)^{(2k+j)}\right| + \left|(C^m)^{(n-j)}\right|$ for all $j \in [1,k-1]$.
7. $B^m_{[j,2k-j]\cup[2k+j,n-j]} \geq C^m_{[j,2k-j]\cup[2k+j,n-j]}$.
8. $B^m_{[j,n-j]} \geq C^m_{[j,n-j]}$ for all $j \in [1,2k]$.

(1 is 6.4 (i), (ii), (iii) and (iv), 2 is 6.4 (v), 3 is 6.4 (vi) and 4 is 6.4 (vii).)

Note that item 8 with $j = 1$ shows B dominates C, which is what we are really after. We reemphasize that the use of the notation $\langle a,b\rangle^j$ as an alternative for $C^{(j)}$ in Lemma 6.4 does not mean that Y_j is of the form $[a,b]$ but rather that a is the smallest element of $[1,n-1]$ for which $c_{i\,j} = 1$ and b is the largest element of $[1,n-1]$ for which $c_{i\,j} = 1$.

PROOF. Let column $B^{(j)}$ be denoted by $\langle x,y\rangle^j$ throughout with $X_j = \{i \in [1,n-1] : b_{i\,j} = 1\}$ and $Y_j = \{i \in [1,n-1] : c_{i\,j} = 1\}$. The proof is by induction. For any $p \geq 1$ let Claim (1, p), Claim (2, p), ..., Claim (11, p) be as follows:

Claim (1, p) $\left|\left(B^{p-1}\right)^{(2i-1)}\right| \geq \left|\left(C^{p-1}\right)^{(n-2i+1)}\right|$ for all $i \in [1,k/2]$.

Claim (2, p) $\left|\left(B^{p-1}\right)^{(n-2i)}\right| \geq \left|\left(C^{p-1}\right)^{(2i)}\right|$ for all $i \in [1,(k-1)/2]$.

Claim (3, p) $\left|\left(B^{p-1}\right)^{(2k+2i-1)}\right| \geq \left|\left(C^{p-1}\right)^{(2k-2i+1)}\right|$ for all $i \in [1,(k+1)/2]$.

Claim (4, p) $\left|\left(B^{p-1}\right)^{(2k+2i)}\right| \geq \left|\left(C^{p-1}\right)^{(2k-2i)}\right|$ for all $i \in [1,k/2]$.

Claim (5, p) $\left|(B^p)^{(j)}\right| \geq \left|(C^p)^{(j)}\right|$ for all $j \in O[1,k-1] \cup \{k\} \cup [2k,3k] \cup E[3k+1,n-2]$ and for all j such that

(a) $j = 2i$, with $i \in [1,(k-1)/2]$ and $\langle a,b\rangle^{2i} \neq \langle 2k-2i, 2k+2i-1\rangle$

or (b) $j = n-2i+1$, with $i \in [1,k/2]$ and $\langle a,b\rangle^{n-2i+1} \neq \langle 2k-2i+1, 2k+2i-2\rangle$

or (c) $j = 2k-2i+1$, with $i \in [1,k/2]$ and $\langle a,b\rangle^{2k-2i+1} \neq \langle 2i, n-2i+1\rangle$ or $\langle a,b\rangle^{2k+2i-1} \neq \langle 2i-1, n-2i\rangle$

or (d) $j = 2k-2i$, with $i \in [1,(k-1)/2]$ and $\langle a,b\rangle^{2k-2i} \neq \langle 2i, n-2i-1\rangle$ or $\langle a,b\rangle^{2k+2i} \neq \langle 2i+1, n-2i\rangle$.

Claim (6, p) $\left|(B^p)^{(j)}\right| + \left|(B^p)^{(n-j)}\right| \geq \left|(C^p)^{(j)}\right| + \left|(C^p)^{(n-j)}\right|$ for j such that

(a) $j = 2k-2i+1$, with $i \in [1,(k+1)/2]$ and $\langle a,b\rangle^{2k-2i+1} = \langle 2i, n-2i+1\rangle$ and $\langle a,b\rangle^{n-j} = \langle a,b\rangle^{2k+2i-1} = \langle 2i-1, n-2i\rangle$

or (b) $j = 2k-2i$, with $i \in [1,k/2]$ and $\langle a,b\rangle^{2k-2i} = \langle 2i, n-2i-1\rangle$ and $\langle a,b\rangle^{n-j} = \langle a,b\rangle^{2k+2i} = \langle 2i+1, n-2i\rangle$.

Claim (7, p) $\left|(B^p)^{(j)}\right| + \left|(B^p)^{(s)}\right| \geq \left|(C^p)^{(j)}\right| + \left|(C^p)^{(s)}\right|$ for $j \in E[2,k-1] \cup O[3k+1,n-1]$ such that $C^{(j)}$ and $C^{(s)}$ satisfy one of the following conditions:

(a) $\langle a,b\rangle^{2i} = \langle 2k-2i, 2k+2i-1\rangle$ and $\langle a,b\rangle^{n-2i} \neq \langle 2k-2i+1, 2k+2i\rangle$ for $i \in [1,(k-1)/2]$

(b) $\langle a,b\rangle^{2i} = \langle 2k-2i, 2k+2i-1\rangle$ and $\langle a,b\rangle^{2k-2i} \neq \langle 2i, n-2i-1\rangle$ for $i \in [1,(k-1)/2]$

(c) $\langle a,b\rangle^{2i} = \langle 2k-2i, 2k+2i-1\rangle$ and $\langle a,b\rangle^{2k+2i} \neq \langle 2i+1, n-2i\rangle$ for $i \in [1,(k-1)/2]$

(d) $\langle a,b\rangle^{n-2i+1} = \langle 2k-2i+1, 2k+2i-2\rangle$ and $\langle a,b\rangle^{2i-1} \neq \langle 2k-2i+2, 2k+2i-1\rangle$ for $i \in [1,k/2]$

(e) $\langle a,b\rangle^{n-2i+1} = \langle 2k-2i+1, 2k+2i-2\rangle$ and $\langle a,b\rangle^{2k-2i+1} \neq \langle 2i, n-2i+1\rangle$ for $i \in [1,k/2]$

(f) $\langle a,b\rangle^{n-2i+1} = \langle 2k-2i+1, 2k+2i-2\rangle$ and $\langle a,b\rangle^{2k+2i-1} \neq$

$\langle 2i-1, n-2i\rangle$ for $i \in [1, k/2]$.

Claim (8, p) $B^p_{[j,n-j]} \geq C^p_{[j,n-j]}$ for all $j \in [k, 2k]$.

Claim (9, p) $\left|(B^p)^{(j)}\right| + \left|(B^p)^{(2k-j)}\right| + \left|(B^p)^{(2k+j)}\right| + \left|(B^p)^{(n-j)}\right| \geq \left|(C^p)^{(j)}\right|$ $+ \left|(C^p)^{(2k-j)}\right| + \left|(C^p)^{(2k+j)}\right| + \left|(C^p)^{(n-j)}\right|$ for all $j \in [1, k-1]$.

Claim (10, p) $B^p_{[j,2k-j]\cup[2k+j,n-j]} \geq C^p_{[j,2k-j]\cup[2k+j,n-j]}$ for all $j \in [1, k-1]$.

Claim (11, p) $B^p_{[j,n-j]} \geq C^p_{[j,n-j]}$ for all $j \in [1, 2k]$.

For $p = 1$, $\left|B^{(j)}\right| \geq \left|C^{(j)}\right|$, $\left|B^{(n-j)}\right| \geq \left|C^{(j)}\right|$ and $1 = \left|I^{(n-j)}\right| = \left|(B^0)^{(n-j)}\right| \geq \left|(C^0)^{(j)}\right| = \left|I^{(j)}\right| = 1$ for all $j \in [1, n-1]$ by Corollaries 3.3 and 4.7. (Here I is the identity matrix of the appropriate dimension.) Hence all claims are true for $p = 1$.

We now assume all claims are true for $p = 1$ to $p = m$ (if $r \in [1, 11]$ and $s \in [1, m]$ we refer to the assumed true Claim (r, s) as inductive hypothesis (r, s)) and aim to show all claims are true for $p = m + 1$.

Proof of Claim (1, m+1). To show that $\left|(B^m)^{(2i-1)}\right| \geq \left|(C^m)^{(n-2i+1)}\right|$ for all $i \in [1, k/2]$ we consider two cases:

(a) $\langle a, b\rangle^{n-2i+1} \neq \langle 2k-2i+1, 2k+2i-2\rangle$;

(b) $\langle a, b\rangle^{n-2i+1} = \langle 2k-2i+1, 2k+2i-2\rangle$.

Case (a). By inductive hypothesis (5, m)(b) and Lemma 6.2 we have

$$\left|(C^m)^{(n-2i+1)}\right| \leq \left|(B^m)^{(n-2i+1)}\right| \leq \left|(B^m)^{(2i-1)}\right|.$$

Case (b).

$$\begin{aligned}
&\left|(B^m)^{(2i-1)}\right| - \left|(C^m)^{(n-2i+1)}\right| \\
&\quad = B^{m-1}_{[2k-2i+2,2k+2i-1]} - C^{m-1}_{[2k-2i+1,2k+2i-2]} \\
&\quad = \left(B^{m-1} - C^{m-1}\right)[2k-2i+2, 2k+2i-2] \\
&\qquad + \left|\left(B^{m-1}\right)^{(2k+2i-1)}\right| - \left|\left(C^{m-1}\right)^{(2k-2i+1)}\right| \\
&\quad \geq 0,
\end{aligned}$$

by inductive hypotheses (11, m−1) and (3, m).

Proof of Claim (2, m+1). To show that $\left|(B^m)^{(n-2i)}\right| \geq \left|(C^m)^{(2i)}\right|$ for all $i \in [1, (k-1)/2]$ we consider two cases:

(a) $\langle a, b\rangle^{2i} \neq \langle 2k-2i, 2k+2i-1\rangle$;

(b) $\langle a, b\rangle^{2i} = \langle 2k-2i, 2k+2i-1\rangle$.

Case (a). By inductive hypothesis (5, m)(a) and Lemma 6.2 we have

$$\left|(C^m)^{(2i)}\right| \leq \left|(B^m)^{(2i)}\right| \leq \left|(B^m)^{(n-2i)}\right|.$$

Case (*b*).

$$\begin{aligned}
&\left|(B^m)^{(n-2i)}\right| - \left|(C^m)^{(2i)}\right| \\
&\quad = B^{m-1}_{[2k-2i+1,2k+2i]} - C^{m-1}_{[2k-2i,2k+2i-1]} \\
&\quad = \left(B^{m-1} - C^{m-1}\right)[2k-2i+1, 2k+2i-1] \\
&\qquad + \left|\left(B^{m-1}\right)^{(2k+2i)}\right| - \left|\left(C^{m-1}\right)^{(2k-2i)}\right| \\
&\quad \geq 0,
\end{aligned}$$

by inductive hypotheses (11, m−1) and (4, m).

Proof of Claim (*3, m+1*). To show that $\left|(B^m)^{(2k+2i-1)}\right| \geq \left|(C^m)^{(2k-2i+1)}\right|$ for all $i \in [1, (k+1)/2]$ we consider two cases:

(a) $\langle a, b\rangle^{2k-2i+1} \neq \langle 2i, n-2i+1\rangle$;
(b) $\langle a, b\rangle^{2k-2i+1} = \langle 2i, n-2i+1\rangle$.

Case (*a*). By inductive hypothesis (5, m)(c) and Lemma 6.2 we have

$$\left|(C^m)^{(2k-2i+1)}\right| \leq \left|(B^m)^{(2k-2i+1)}\right| \leq \left|(B^m)^{(2k+2i-1)}\right|.$$

Case (*b*).

$$\begin{aligned}
&\left|(B^m)^{(2k+2i-1)}\right| - \left|(C^m)^{(2k-2i+1)}\right| \\
&\quad = B^{m-1}_{[2i-1,n-2i]} - C^{m-1}_{[2i,n-2i+1]} \\
&\quad = \left(B^{m-1} - C^{m-1}\right)[2i, n-2i] + \left|\left(B^{m-1}\right)^{(2i-1)}\right| - \left|\left(C^{m-1}\right)^{(n-2i+1)}\right| \\
&\quad \geq 0,
\end{aligned}$$

by inductive hypotheses (11, m−1) and (1, m) provided $i \neq (k+1)/2$. If $i = (k+1)/2$ (whence i is odd), then

$$\begin{aligned}
\left|\left(B^{m-1}\right)^{(2k+2i-1)}\right| - \left|\left(C^{m-1}\right)^{(2k-2i+1)}\right| &= \left|\left(B^{m-1}\right)^{(3k)}\right| - \left|\left(C^{m-1}\right)^{(k)}\right| \\
&= \left|\left(B^{m-1}\right)^{(k)}\right| - \left|\left(C^{m-1}\right)^{(k)}\right| \\
&\geq 0,
\end{aligned}$$

by Lemma 6.2 and inductive hypothesis (5, m−1).

Proof of Claim (*4, m+1*). To show that $\left|(B^m)^{(2k+2i)}\right| \geq \left|(C^m)^{(2k-2i)}\right|$ for all $i \in [1, k/2]$ we consider two cases:

(a) $\langle a, b\rangle^{2k-2i} \neq \langle 2i, n-2i-1\rangle$;
(b) $\langle a, b\rangle^{2k-2i} = \langle 2i, n-2i-1\rangle$.

Case (*a*). By inductive hypothesis (5, m)(d) and Lemma 6.2 we have

$$\left|(C^m)^{(2k-2i)}\right| \leq \left|(B^m)^{(2k-2i)}\right| \leq \left|(B^m)^{(2k+2i)}\right|.$$

Case (b).

$$\begin{aligned}
&\left|(B^m)^{(2k+2i)}\right| - \left|(C^m)^{(2k-2i)}\right| \\
&\quad = B^{m-1}_{[2i+1,n-2i]} - C^{m-1}_{[2i,n-2i-1]} \\
&\quad = \left(B^{m-1} - C^{m-1}\right)[2i+1, n-2i-1] + \left|\left(B^{m-1}\right)^{(n-2i)}\right| - \left|\left(C^{m-1}\right)^{(2i)}\right| \\
&\quad \geq 0,
\end{aligned}$$

by inductive hypotheses (11, m−1) and (2, m) provided $i \neq k/2$. If $i = k/2$ (whence i is even), then

$$\begin{aligned}
\left|\left(B^{m-1}\right)^{(n-2i)}\right| - \left|\left(C^{m-1}\right)^{(2i)}\right| &= \left|\left(B^{m-1}\right)^{(3k)}\right| - \left|\left(C^{m-1}\right)^{(k)}\right| \\
&\geq 0,
\end{aligned}$$

by inductive hypothesis (4, m).

Note that as a consequence of Claims (1, m+1), (2, m+1), (3, m+1), (4, m+1), inductive hypothesis (5, m) and Lemma 6.2 and Corollary 5.6, we have

Corollary (6.5, m). *For any* $j \in [2, 2k]$

$$\left|(B^m)^{(j)}\right| \geq \left|(C^m)^{(j-1)}\right| \text{ and } \left|(B^m)^{(j)}\right| \geq \left|(C^m)^{(n-j+1)}\right|$$

and

$$\left|(B^m)^{(n-j)}\right| \geq \left|(C^m)^{(n-j+1)}\right| \text{ and } \left|(B^m)^{(n-j)}\right| \geq \left|(C^m)^{(j-1)}\right|.$$

Proof. To see this

(i) Let $i \in [2, k/2]$. Then

$$\begin{aligned}
\left|(B^m)^{(2i-1)}\right| &\geq \left|(B^m)^{(n-2i+1)}\right| && \text{(by Lemma 6.2)} \\
&\geq \left|(B^m)^{(n-2i+2)}\right| && \text{(by Corollary 5.6)} \\
&\geq \begin{cases} \left|(C^m)^{(n-2i+2)}\right| & \text{(by ind. hyp. (5, m)) or} \\ \left|(C^m)^{(2i-2)}\right| & \text{(by Claim (2, m+1)).} \end{cases}
\end{aligned}$$

(ii) Let $i \in [1, (k-1)/2]$. Then

$$\begin{aligned}
\left|(B^m)^{(n-2i)}\right| &\geq \left|(B^m)^{(2i)}\right| && \text{(by Lemma 6.2)} \\
&\geq \left|(B^m)^{(2i-1)}\right| && \text{(by Corollary 5.6)} \\
&\geq \begin{cases} \left|(C^m)^{(2i-1)}\right| & \text{(by ind. hyp. (5, m)) or} \\ \left|(C^m)^{(n-2i+1)}\right| & \text{(by Claim (1, m+1)).} \end{cases}
\end{aligned}$$

(iii) Let $i \in [1, k/2]$. Then

$$\begin{aligned}
\left|(B^m)^{(2k+2i-1)}\right| &\geq \left|(B^m)^{(2k-2i+1)}\right| && \text{(by Lemma 6.2)} \\
&\geq \left|(B^m)^{(2k+2i)}\right| && \text{(by Lemma 6.2)} \\
&\geq \begin{cases} \left|(C^m)^{(2k+2i)}\right| & \text{(by ind. hyp. (5, m)) or} \\ \left|(C^m)^{(2k-2i)}\right| & \text{(by Claim (4, m+1)).} \end{cases}
\end{aligned}$$

(iv) Let $i \in [1, (k-1)/2]$. Then

$$\begin{aligned}\left|(B^m)^{(2k+2i)}\right| &\geq \left|(B^m)^{(2k-2i)}\right| \quad \text{(by Lemma 6.2)}\\ &\geq \left|(B^m)^{(2k+2i+1)}\right| \quad \text{(by Lemma 6.2)}\\ &\geq \begin{cases} \left|(C^m)^{(2k+2i+1)}\right| & \text{(by ind. hyp. (5, m)) or}\\ \left|(C^m)^{(2k-2i-1)}\right| & \text{(by Claim (3, m+1)).}\end{cases}\end{aligned}$$

(v) Let $j = 2k$. Then

$$\begin{aligned}\left|(B^m)^{(2k)}\right| &\geq \left|(B^m)^{(2k+1)}\right| \quad \text{(by Corollary 5.6)}\\ &\geq \begin{cases} \left|(C^m)^{(2k-1)}\right| & \text{(by Claim (3, m+1)) or}\\ \left|(C^m)^{(2k+1)}\right| & \text{(by ind. hyp. (5, m)).}\end{cases}\end{aligned}$$

(vi) Let $j = k$. Then

$$\begin{aligned}\left|(B^m)^{(k)}\right| &= \left|(B^m)^{(3k)}\right|\\ &\geq \left|(C^m)^{(3k)}\right| \quad \text{(by ind. hyp. (5, m))}\\ &\geq \left|(C^m)^{(3k+1)}\right| \quad \text{(by Corollary 5.6)}\end{aligned}$$

and

$$\begin{aligned}\left|(B^m)^{(3k)}\right| &= \left|(B^m)^{(k)}\right|\\ &\geq \left|(C^m)^{(k)}\right| \quad \text{(by ind. hyp. (5, m))}\\ &\geq \left|(C^m)^{(k-1)}\right| \quad \text{(by Corollary 5.6).}\end{aligned}$$

This completes the proof of Corollary (6.5, m). □

Proof of Claims (5, m+1) and (7, m+1). We first remark that it is easy to see that Claim (5, p) is equivalent to the following:

Claim (5′, p) $\quad \left|(B^p)^{(j)}\right| \geq \left|(C^p)^{(j)}\right| \quad$ for all $j \in [1, n-1]$

such that

(*a*) $j \notin [k+1, 2k-1]$ and $\langle a,b\rangle^j = \langle x,y\rangle^j$

or (*b*) $j = 2i-1$, with $i \in [1, k/2]$ and $\langle a,b\rangle^{2i-1} \neq (\langle x,y\rangle^{2i-1} =)\langle 2k-2i+2, 2k+2i-1\rangle$

or (*c*) $j = n-2i$, with $j \in [1, (k-1)/2]$ and $\langle a,b\rangle^{n-2i} \neq (\langle x,y\rangle^{n-2i} =)\langle 2k-2i+1, 2k+2i\rangle$

or (*d*) $j = 2i$, with $i \in [1, (k-1)/2]$ and $\langle a,b\rangle^{2i} \neq (\langle x,y\rangle^{2i} =)\langle 2k-2i+2,$

$2k+2i+1\rangle$ and $\langle a,b\rangle^{2i} \neq \langle 2k-2i, 2k+2i-1\rangle$

or (e) $j = n-2i+1$, with $i \in [1, k/2]$ and $\langle a,b\rangle^{n-2i+1} \neq (\langle x,y\rangle^{n-2i+1} =)\langle 2k-$ $2i+3, 2k+2i\rangle$ and $\langle a,b\rangle^{n-2i+1} \neq \langle 2k-2i+1, 2k+2i-2\rangle$

or (f) $j = 2k+2i-1$, with $i \in [1, (k+1)/2]$ and $\langle a,b\rangle^{2k+2i-1} \neq (\langle x,y\rangle^{2k+2i-1} =)\langle 2i-1, n-2i\rangle$

or (g) $j = 2k+2i$, with $i \in [1, k/2]$ and $\langle a,b\rangle^{2k+2i} \neq (\langle x,y\rangle^{2k+2i} =)\langle 2i+$ $1, n-2i\rangle$

or (h) $j = 2k-2i+1$, with $i \in [1, (k+1)/2]$ and $\langle a,b\rangle^{2k-2i+1} \neq (\langle x,y\rangle^{2k-2i+1} =)\langle 2i, n-2i+1\rangle$

or (h') $j = 2k-2i+1$, with $i \in [1, k/2]$ and $\langle a,b\rangle^{2k-2i+1} = \langle 2i, n-2i+1\rangle$ and $\langle a,b\rangle^{2k+2i-1} \neq \langle 2i-1, n-2i\rangle$

or (i) $j = 2k-2i$, with $i \in [1, k/2]$ and $\langle a,b\rangle^{2k-2i} \neq (\langle x,y\rangle^{2k-2i} =)\langle 2i, n-2i-1\rangle$

or (i') $j = 2k-2i$, with $i \in [1, (k-1)/2]$ and $\langle a,b\rangle^{2k-2i} = \langle 2i, n-2i-1\rangle$ and $\langle a,b\rangle^{2k+2i} \neq \langle 2i+1, n-2i\rangle$.

Thus we will establish Claim (5′, m+1).

In fact we will only deal directly with the cases (a), (h′), and (i′) where specifically we know $\langle a,b\rangle^j = \langle x,y\rangle^j$. The multitudinous cases remaining to be proved in Claim (5′, m+1), and the even greater number of cases remaining in Claim (7, m+1) will be handled using Corollary 5.13 and Lemma (6.7, m) below, for which, in turn, we devise a number of algorithms to generate proofs in all possible cases.

Direct proofs of (a), (h′) and (i′) of Claim (5′, m+1).

(a) Let $j \notin [k+1, 2k-1]$ be such that $\langle a,b\rangle^j = \langle x,y\rangle^j$. For $j = 2i-1$ and $1 \leq i \leq k/2$ (that is, $j \in O[1, k-1]$) we have $\langle a,b\rangle^j = \langle x,y\rangle^j = \langle 2k-2i+2, 2k+2i-1\rangle$, and so

$$\begin{aligned}\left|\left(B^{m+1}\right)^{(j)}\right| - \left|\left(C^{m+1}\right)^{(j)}\right| &= (B^m - C^m)[2k-2i+2, 2k+2i-2] \\ &\quad + \left|(B^m)^{(2k+2i-1)}\right| - \left|(C^m)^{(2k+2i-1)}\right| \\ &\geq 0,\end{aligned}$$

by inductive hypotheses (11, m) and (5, m).

Similar proofs that $\left|\left(B^{m+1}\right)^{(j)}\right| - \left|\left(C^{m+1}\right)^{(j)}\right| \geq 0$ for identical reasons hold in the cases $j = 2i$, $1 \leq i \leq (k-1)/2$; $j = k$, k odd; $j = k$, k even; $j = 2k+2i-1$, $1 \leq i \leq (k+1)/2$; $j = 2k+2i$, $1 \leq i \leq k/2$; $j = n-2i$, $1 \leq i \leq (k-1)/2$ and $j = n-2i+1$, $1 \leq i \leq k/2$. This covers every case except for the case $j = 2k$ which is even simpler as it merely relies on inductive hypothesis (11, m).

(h′) Let $j = 2k-2i+1$, with $i \in [1, k/2]$ and $\langle a,b\rangle^{2k-2i+1} = \langle x,y\rangle^{2k-2i+1} = \langle 2i, n-2i+1\rangle$ but $\langle a,b\rangle^{2k+2i-1} \neq (\langle x,y\rangle^{2k+2i-1} =)\langle 2i-1, n-2i\rangle$. Then

$$\left|\left(B^{m+1}\right)^{(j)}\right| - \left|\left(C^{m+1}\right)^{(j)}\right| = (B^m - C^m)[2i, n-2i+1].$$

Case (1). When $\langle a,b\rangle^{n-2i+1} \neq \langle 2k-2i+1, 2k+2i-2\rangle$.

$$\begin{aligned}\left|\left(B^{m+1}\right)^{(j)}\right| - \left|\left(C^{m+1}\right)^{(j)}\right| &= (B^m - C^m)[2i, n-2i] \\ &\quad + \left|(B^m)^{(n-2i+1)}\right| - \left|(C^m)^{(n-2i+1)}\right| \\ &\geq 0\end{aligned}$$

by inductive hypotheses (11, m) and (5′, m)(e).

Case (2). When $\langle a,b\rangle^{n-2i+1} = \langle 2k-2i+1, 2k+2i-2\rangle$.

$$\begin{aligned}\left|\left(B^{m+1}\right)^{(j)}\right| - \left|\left(C^{m+1}\right)^{(j)}\right| &= (B^m - C^m)([2i, 2k-2i] \cup [2k+2i, n-2i]) \\ &\quad + (B^m - C^m)[2k-2i+2, 2k+2i-2] \\ &\quad + \left|(B^m)^{(n-2i+1)}\right| + \left|(B^m)^{(2k+2i-1)}\right| \\ &\quad - \left(\left|(C^m)^{(n-2i+1)}\right| + \left|(C^m)^{(2k+2i-1)}\right|\right) \\ &\quad + \left|(B^m)^{(2k-2i+1)}\right| - \left|(C^m)^{(2k-2i+1)}\right| \\ &\geq 0\end{aligned}$$

by inductive hypotheses (10, m), (11, m), (7, m)(f) and (5′, m)(h′).

(i′) Let $j = 2k-2i$ with $i \in [1, (k-1)/2]$ and $\langle a,b\rangle^{2k-2i} = \langle x,y\rangle^{2k-2i} = \langle 2i, n-2i-1\rangle$ but $\langle a,b\rangle^{2k+2i} \neq (\langle x,y\rangle^{2k+2i} =)\langle 2i+1, n-2i\rangle$. Then

$$\left|\left(B^{m+1}\right)^{(j)}\right| - \left|\left(C^{m+1}\right)^{(j)}\right| = (B^m - C^m)[2i, n-2i-1].$$

Case (1). When $\langle a,b\rangle^{2i} \neq \langle 2k-2i, 2k+2i-1\rangle$.

$$\begin{aligned}\left|\left(B^{m+1}\right)^{(j)}\right| - \left|\left(C^{m+1}\right)^{(j)}\right| &= (B^m - C^m)[2i+1, n-2i-1] \\ &\quad + \left|(B^m)^{(2i)}\right| - \left|(C^m)^{(2i)}\right| \\ &\geq 0\end{aligned}$$

by inductive hypotheses (11, m) and (5′, m)(d).

Case (2). When $\langle a,b\rangle^{2i} = \langle 2k-2i, 2k+2i-1\rangle$.

$$\begin{aligned}\left|\left(B^{m+1}\right)^{(j)}\right| - \left|\left(C^{m+1}\right)^{(j)}\right| &= (B^m - C^m)([2i+1, 2k-2i-1] \\ &\quad \cup [2k+2i+1, n-2i-1]) \\ &\quad + (B^m - C^m)[2k-2i+1, 2k+2i-1] \\ &\quad + \left|(B^m)^{(2i)}\right| + \left|(B^m)^{(2k+2i)}\right| \\ &\quad - \left(\left|(C^m)^{(2i)}\right| + \left|(C^m)^{(2k+2i)}\right|\right) \\ &\quad + \left|(B^m)^{(2k-2i)}\right| - \left|(C^m)^{(2k-2i)}\right| \\ &\geq 0\end{aligned}$$

by inductive hypotheses (10, m), (11, m), (7, m)(c) and (5′, m)(i′).

We now present a definition and an accompanying lemma designed to greatly simplify the proof of the remainder of Claim (5′, m+1) and of Claim (7, m+1).

DEFINITION 6.6. For any non-empty subset W of $[1, n-1]$ let $J(W) = \{j \in [0, 2k-1] : [2k-j, 2k+j] \subseteq W\}$, then the *centre cut* $c(W)$ of W is given by

$$c(W) = \begin{cases} \emptyset, & \text{if } J = \emptyset \\ [2k-j, 2k+j] \text{ where } j = \max\ J(W), & \text{if } J \neq \emptyset. \end{cases}$$

Further, if $J = \emptyset$, which is true if and only if $2k \notin W$, then the *centre span* $s(W)$ of W is given by

$$s(W) = W \cap \{2k-j, 2k+j\} \text{ where } j = 1 + \max\ J([1, n-1] \setminus W).$$

Finally, if $J \neq \emptyset$, $s(W) = \{2k\}$. (So $c(W)$ is the " largest" subinterval of W which is symmetric about $2k$, and $s(W)$ is the set of elements of W which are " closest" to $2k$.) We also make use of the set $t(W)$ given by $t(W) = s(W) \cup \{n-j : j \in s(W)\}$.

LEMMA (6.7, M).
(I) $\left|\left(B^{m+1}\right)^{(j)}\right| = B^m_{X_j} \geq C^m_W$ *for*
(i) $j = 2i-1$, *with* $i \in [1, k/2]$ *and* $W \in (T_{2i-1} \setminus \{X_{2i-1}\}) \cup \{[2k-2i+1, 2k+2i-2]\}$
(ii) $j = n-2i$, *with* $i \in [1, (k-1)/2]$ *and* $W \in (T_{n-2i} \setminus \{X_{n-2i}\}) \cup \{[2k-2i, 2k+2i-1]\}$
(iii) $j = 2i$, *with* $i \in [1, (k-1)/2]$ *and* $W \in T_{2i} \setminus \{X_{2i}\}$
(iv) $j = n-2i+1$, *with* $i \in [1, k/2]$ *and* $W \in T_{n-2i+1} \setminus \{X_{n-2i+1}\}$
(v) $j = 2k+2i-1$, *with* $i \in [1, (k+1)/2]$ *and* $W \in T_{2k+2i-1} \setminus \{X_{2k+2i-1}\}$
(vi) $j = 2k+2i$, *with* $i \in [1, k/2]$ *and* $W \in T_{2k+2i} \setminus \{X_{2k+2i}\}$
(vii) $j = 2k-2i+1$, *with* $i \in [1, (k+1)/2]$ *and* $W \in T'_{2k-2i+1}$
(viii) $j = 2k-2i$, *with* $i \in [1, k/2]$ *and* $W \in T'_{2k-2i}$.
(II) $B^m_Z \geq C^m_W$ *for*
(i) $Z = X_{2i}$, *with* $i \in [1, (k-1)/2]$ *and* $W \in T'_{n-2i}$
(ii) $Z = X_{n-2i+1}$, *with* $i \in [1, k/2]$ *and* $W \in T'_{2i-1}$
(iii) $Z = [2i+2, n-2i+1]$, *with* $i \in [1, (k-1)/2]$ *and* $W \in T'_{2k+2i} \setminus \{[2i-1, n-2i-2], [2i-1, 2i] \cup [2i+3, n-2i], [1, n-1] \cap ([2i+1, n-2i-2] \cup [n-2i+1, n-2i+2])\}$
(iv) $Z = [2i-1, n-2i-2]$, *with* $i \in [1, (k-1)/2]$ *and* $W \in T'_{2k-2i} \setminus \{[2i+2, n-2i+1], [2i, n-2i-3] \cup [n-2i, n-2i+1]\}$
(v) $Z = [2i+1, n-2i+2]$, *with* $i \in [2, k/2]$ *and* $W \in T'_{2k-2i+1} \setminus \{[2i-2, n-2i-1], [2i-2, 2i-1] \cup [2i+2, n-2i+1]\}$
(vi) $Z = [2i-2, n-2i-1]$, *with* $i \in [2, k/2]$ *and* $W \in T'_{2k+2i-1} \setminus \{[2i+1, n-2i+2], [2i-1, n-2i-2] \cup [n-2i+1, n-2i+2], [2i-3, 2i-2] \cup [2i+1, n-2i]\}$.

PROOF. In each case we wish to show $B^m_Z \geq C^m_W$; that is, $B^m_Z - C^m_W \geq 0$, for specified subsets of Z and W of $[1, n-1]$.

The proof of this result, which is central to the paper, is extremely repetitive. In fact it is possible to set up a single algorithm which produces a proof of every case of the result, but the algorithm itself is overly complicated in a large number of the cases. A better solution is to use two much simpler algorithms, the first for handling the vast majority of cases, and the second for handling the remaining cases.

ALGORITHM 6.8. This produces a proof of all the above cases with the exception of the following:

I(i) when $W = [2k - 2i + 1, 2k + 2i - 2]$
I(ii) when $W = [2k - 2i, 2k + 2i - 1]$
I(iii) when $W = [2k - 2i, 2k + 2i - 3] \cup [2k + 2i, 2k + 2i + 1]$
I(iv) when $W = [2k - 2i + 1, 2k + 2i - 4] \cup [2k + 2i - 1, 2k + 2i]$ and $i > 1$
II(i) when $W = [2k - 2i - 1, 2k + 2i - 2]$
II(i) when $W = [2k - 2i - 1, 2k - 2i] \cup [2k - 2i + 3, 2k + 2i]$
II(ii) when $W = [2k - 2i, 2k + 2i - 3]$
II(ii) when $W = [2k - 2i, 2k - 2i + 1] \cup [2k - 2i + 4, 2k + 2i - 1]$ and $i > 1$.

Step 1. (Complete exploitation of inductive hypothesis (11, m).)

Consider $B_Z^m - C_W^m$. In every case it can be observed that $c(W) \subseteq c(Z)$ whence $c(W) \subseteq Z \cap W$. By inductive hypothesis (11, m) we have $B_{c(W)}^m \geq C_{c(W)}^m$, thus

$$\begin{aligned} B_Z^m - C_W^m &= B_Z^m - B_{c(W)}^m + B_{c(W)}^m - C_W^m \\ &\geq B_Z^m - B_{c(W)}^m + C_{c(W)}^m - C_W^m \\ &= B_{Z_1}^m - C_{W_1}^m, \end{aligned}$$

where $Z_1 = Z \setminus c(W)$ and $W_1 = W \setminus c(W)$.

To prove the results it now suffices to show $B_{Z_1}^m \geq C_{W_1}^m$. We remark that $Z = Z_1$ and $W = W_1$ if $2k \notin W$, and that in the vast majority of cases Z_1 is the union of two disjoint intervals and $s(Z_1)$ is a two element set while W_1 is a single interval with $s(W_1)$ a one element set. (In any case, in all cases $s(W_1)$ is a one element set, and Z_1 and W_1 are either single intervals or unions of two disjoint intervals.) Further, unless $2k \notin W$, in each case $t(Z_1) = t(W_1)$.

Step 2. (To be used only in the case where $t(Z_1) = t(W_1)$ otherwise, set $Z_2 = Z_1$ and $W_2 = W_1$ and go to Step 3. Step 2 is a simple use of whichever of inductive hypothesis (5′, m) or Claims (1, 2, 3 or 4, m) are appropriate to eliminate the "centre most" element of W_1.) Consider $B_{Z_1}^m - C_{W_1}^m$, and let j be the single element in $s(W_1) \subseteq t(Z_1)$. Recall that $j \neq 2k$.

(i) If $j \in [2k+1, 3k]$, observe that also $j \in Z_1$, and that $B_{\{j\}}^m \geq C_{\{j\}}^m$ by inductive hypothesis (5′, m)(a), (f) and (g). Now

$$\begin{aligned} B_{Z_1}^m - C_{W_1}^m &= B_{Z_1}^m - B_{\{j\}}^m + B_{\{j\}}^m - C_{W_1}^m \\ &\geq B_{Z_1}^m - B_{\{j\}}^m + C_{\{j\}}^m - C_{W_1}^m \\ &= B_{Z_2}^m - C_{W_2}^m, \end{aligned}$$

where $Z_2 = Z_1 \setminus \{j\}$ and $W_2 = W_1 \setminus \{j\}$.

(ii) If $j \in [k, 2k-1]$, observe that $n - j \in Z_1$, and that $B_{\{n-j\}}^m \geq C_{\{j\}}^m$ by Claim (3, m+1) (if $j \in O[k, 2k-1]$) or Claim (4, m+1) (if $j \in E[k, 2k-2]$). Now

$$B_{Z_1}^m - C_{W_1}^m \geq B_{Z_2}^m - C_{W_2}^m,$$

where $Z_2 = Z_1 \setminus \{n-j\}$ and $W_2 = W_1 \setminus \{j\}$.

(iii) If $j \in O[1,k-1] \cup E[3k+1,n-2]$, observe that $j \in Z_1$, and that $B^m_{\{j\}} \geq C^m_{\{j\}}$ by inductive hypothesis $(5', \mathrm{m})$(a), (b) and (c). Now

$$B^m_{Z_1} - C^m_{W_1} \geq B^m_{Z_2} - C^m_{W_2},$$

where $Z_2 = Z_1 \setminus \{j\}$ and $W_2 = W_1 \setminus \{j\}$.

(iv) If $j \in E[2,k-1] \cup O[3k+1,n-1]$, observe that $n-j \in Z_1$, and that $B^m_{\{n-j\}} \geq C^m_{\{j\}}$ by Claim (2, m+1) (if $j \in E[2,k-1]$) or Claim (1, m+1) (if $j \in O[3k+1,n-1]$). Now

$$B^m_{Z_1} - C^m_{W_1} \geq B^m_{Z_2} - C^m_{W_2},$$

where $Z_2 = Z_1 \setminus \{n-j\}$ and $W_2 = W_1 \setminus \{j\}$.

The problem now reduces to demonstrating $B^m_{Z_2} \geq C^m_{W_2}$. Furthermore, in each case Algorithm 6.8 claims to deal with, the following properties of Z_2 and W_2 may be verified

(i) $W_2 = \emptyset$ and we are finished.

Otherwise we have

(ii) $t(Z_2) \neq t(W_2)$ and in fact if $j \in t(Z_2) \cap [1,2k]$ and $j' \in t(W_2) \cap [1,2k]$, $j' < j$, (that is, the element of Z_2 closest to $2k$ is closer to $2k$ than the element of W_2 closest to $2k$), and

(iii) Z_2 is either a single interval or a union of two disjoint intervals and in the latter case the element closest to $2k$ in one of these intervals is one closer to $2k$ than the element of the other interval which is closest to $2k$ and

(iv) W_2 is either a single interval or a union of two disjoint intervals and in each case the element of W_2 furthest from $2k$ is further from $2k$ than the element of Z_2 furthest from $2k$, or there is one less element in W_2 than in Z_2 and the element of W_2 furthest from $2k$ is at least as far from $2k$ as the element of Z_2 furthest from $2k$. (It is these latter " distance from $2k$" properties which do not hold in the cases not handled by Algorithm 6.8.)

In each case it is a simple matter to verify that full exploitation of Corollaries 5.7 and (48, m) as described in the rest of Algorithm 6.8 completes the proof of the lemma. The remaining step(s) only apply if $W_2 \neq \emptyset$.

Step n+1. (For $2 \leq n < \overline{n}$, where $\overline{n}$ is determined as the smallest integer for which $W_{\overline{n}} = \emptyset$).

Let $j' = \max\ s(W_n)$ and $j = \max\ s(Z_n)$ (notice that $s(W_2) = \{j'\}$ and $s(Z_2) = \{j\}$).

(i) If $j' < 2k$ and $j \leq 2k$, $j'+1 \leq j \leq 2k$. Then if $I = \{i \in [0, j'-1] : j'-i \in W_n \text{ and } j-i \in Z_n\}$, $0 \in I$, and for each $i \in I$, $1 \leq j'-i < j'-i+1 \leq j-i \leq 2k$ and

$$\begin{aligned} B^m_{\{j-i\}} &\geq B^m_{\{j'-i+1\}} && \text{(by Corollary 5.7)} \\ &\geq C^m_{\{j'-i\}} && \text{(by Corollary (6.5, m))}. \end{aligned}$$

So if we set $Z_{n+1} = Z_n \setminus \{j-i : i \in I\}$ and $W_{n+1} = W_n \setminus \{j'-i : i \in I\}$, then

$$B^m_{Z_n} - C^m_{W_n} \geq B^m_{Z_{n+1}} - C^m_{W_{n+1}}.$$

(ii) If $j' > 2k$ and $j \geq 2k$, $2k \leq j \leq j' - 1$. Then if $I = \{i \in [0, n - j' - 1] : j' + i \in W_n \text{ and } j + i \in Z_n\}$, $0 \in I$, and for each $i \in I$, $2k \leq j + i \leq j' + i - 1 < j' + i \leq n - 1$, and

$$\begin{aligned} B^m_{\{j+i\}} &\geq B^m_{\{j'+i-1\}} && \text{(by Corollary 5.7)} \\ &\geq C^m_{\{j'+i\}} && \text{(by Corollary (6.5, m)).} \end{aligned}$$

So if we set $Z_{n+1} = Z_n \setminus \{j + i : i \in I\}$ and $W_{n+1} = W_n \setminus \{j' + i : i \in I\}$ then

$$B^m_{Z_n} - C^m_{W_n} \geq B^m_{Z_{n+1}} - C^m_{W_{n+1}}.$$

(iii) If $j' < 2k$ and $j > 2k$, $n - j' > j$ so $2k < j \leq n - j' - 1$. Then if $I = \{i \in [0, j' - 1] : j' - i \in W_n \text{ and } j + i \in Z_n\}$, $0 \in I$, and for each $i \in I$, $2k \leq j + i \leq n - j' + i - 1 < n - (j' - i) \leq n - 1$ and

$$\begin{aligned} B^m_{\{j+i\}} &\geq B^m_{\{n-j'+i-1\}} && \text{(by Corollary 5.7)} \\ &\geq C^m_{\{j'-i\}} && \text{(by Corollary (6.5, m)).} \end{aligned}$$

So if we set $Z_{n+1} = Z_n \setminus \{j + i : i \in I\}$ and $W_{n+1} = W_n \setminus \{j' - i : i \in I\}$ then

$$B^m_{Z_n} - C^m_{W_n} \geq B^m_{Z_{n+1}} - C^m_{W_{n+1}}.$$

(iv) If $j' > 2k$ and $j < 2k$, $j > n - j'$ so $n - j' + 1 \leq j < 2k$. Then if $I = \{i \in [0, n - j' - 1] : j' + i \in W_n \text{ and } j - i \in Z_n\}$, $0 \in I$, and for each $i \in I$, $1 \leq n - (j' + i) < n - j' - i + 1 \leq j - i < 2k$ and

$$\begin{aligned} B^m_{\{j-i\}} &\geq B^m_{\{n-j'-i+1\}} && \text{(by Corollary 5.7)} \\ &\geq C^m_{\{j'+i\}} && \text{(by Corollary (6.5, m)).} \end{aligned}$$

So if we set $Z_{n+1} = Z_n \setminus \{j - i : i \in I\}$ and $W_{n+1} = W_n \setminus \{j' + i : i \in I\}$ then

$$B^m_{Z_n} - C^m_{W_n} \geq B^m_{Z_{n+1}} - C^m_{W_{n+1}}.$$

We illustrate the use of Algorithm 6.8 in a couple of the more unusual cases (that is, those far removed from the very straightforward and typical instances when Z_1 is the union of two disjoint intervals with $s(Z_1)$ a two element set, and W_1 is a single interval).

EXAMPLE 6.9. In the proof of I(vii) we consider the case where $W = [1, n - 1] \cap ([2i - 2, 2i - 1] \cup [2i + 2, n - 2i + 1])$ and $Z = X_{2k-2i+1} = [2i, n - 2i + 1]$, for $i \in [1, (k + 1)/2]$.

After Step 1 we found that $Z_1 = Z \setminus [2i+2, n-2i-2] = [2i, 2i+1] \cup [n-2i-1, n-2i+1]$ and $W_1 = W \setminus [2i+2, n-2i-2] = [1, n-1] \cap ([2i-2, 2i-1] \cup [n-2i-1, n-2i+1])$.

For the application of Step 2,

$$j = n - 2i - 1 \in \begin{cases} O[2k + 1, 3k], & \text{if } i \in [(k - 1)/2, (k + 1)/2] \\ O[3k + 1, n - 1], & \text{if } i \in [1, (k - 2)/2]. \end{cases}$$

Thus $W_2 = [1, n - 1] \cap ([2i - 2, 2i - 1] \cup [n - 2i, n - 2i + 1])$ and

$$Z_2 = \begin{cases} [2i, 2i + 1] \cup [n - 2i, n - 2i + 1], & \text{if } i \in [(k - 1)/2, (k + 1)/2] \\ \{2i\} \cup [n - 2i - 1, n - 2i + 1], & \text{if } i \in [1, (k - 1)/2]. \end{cases}$$

For the application of Step 3, $j' = \max\ s(W_2) = n - 2i > 2k$ and

$$j = \max\ s(Z_2) = \begin{cases} 2i+1 < 2k, & \text{if } i \in [(k-1)/2, (k+1)/2] \\ n-2i-1 \geq 2k, & \text{if } i \in [1, (k-2)/2]. \end{cases}$$

It follows that $I = [0,1]$ in each case, $W_3 = W_2 \setminus [n-2i, n-2i+1] = [1, n-1] \cap [2i-2, 2i-1]$ in each case, and

$$Z_3 = \begin{cases} Z_2 \setminus [2i, 2i+1] = [n-2i, n-2i+1], & \text{if } i \in [(k-1)/2, (k+1)/2] \\ Z_2 \setminus [n-2i-1, n-2i] = \{2i, n-2i+1\}, & \text{if } i \in [1, (k-2)/2]. \end{cases}$$

For the application of Step 4, $j' = \max\ s(W_3) = 2i - 1 < 2k$ and

$$j = \max\ s(Z_3) = \begin{cases} n-2i > 2k, & \text{if } i \in [(k-1)/2, (k+1)/2] \\ 2i \leq 2k, & \text{if } i \in [1, (k-2)/2]. \end{cases}$$

Now

$$I = \begin{cases} [0,1], & \text{if } i \in [(k-1)/2, (k+1)/2] \\ \{0\}, & \text{if } i \in [1, (k-2)/2], \end{cases}$$

so

$$W_4 = \begin{cases} W_3 \setminus [2i-2, 2i-1] = \emptyset, & \text{if } i \in [(k-1)/2, (k+1)/2] \\ W_3 \setminus \{2i-1\} = & \begin{cases} \{2i-2\}, & \text{if } i \in [2, (k-2)/2] \\ \emptyset, & \text{if } i = 1 \end{cases} \end{cases}$$

and

$$Z_4 = \begin{cases} Z_3 \setminus [n-2i, n-2i+1] = \emptyset, & \text{if } i \in [(k-1)/2, (k+1)/2] \\ Z_3 \setminus \{2i\} = \{n-2i+1\}, & \text{if } i \in [1, (k-2)/2]. \end{cases}$$

We are now finished except in the case where $i \in [2, (k-2)/2]$ when $W_4 = \{2i-2\}$ and $Z_4 = \{n-2i+1\}$. Here, for the application of Step 5, $j' = 2i-2$, $j = n-2i+1$, $I = \{0\}$, $W_5 = Z_5 = \emptyset$ and the algorithm completes the task.

EXAMPLE 6.10. In the proof of I(v) we consider the case where

$$W = [4i-1, n-1] \text{ and } Z = X_{2k+2i-1} = [2i-1, n-2i],$$

for $i \in [1, (k+1)/2]$.

After Step 1 we found that $Z_1 = Z \setminus [4i-1, n-4i+1] = [2i-1, 4i-2] \cup [n-4i+2, n-2i]$ and $W_1 = W \setminus [4i-1, n-4i+1] = [n-4i+2, n-1]$.

This example will illustrate the "typical" behaviour if $i \in [2, k/2]$ (and is included because of this). For the case $i = 1$, however, we have

$$Z_1 = \{1, 2\} \cup \{n-2\} \text{ and } W_1 = \{n-2, n-1\},$$

which, since $1 \in Z_1$ and $n-1 \in W_1$, would normally be worrying. (Except that W_1 has one less element than Z_1.)

For the application of Step 2,

$$j = n - 4i + 2 \in \begin{cases} E[2k+2, 3k], & \text{for } i \in [(k+2)/4, (k+1)/2] \\ E[3k+1, n-2], & \text{for } i \in [1, (k+1)/4]. \end{cases}$$

Thus, in either case, $W_2 = [n-4i+3, n-1]$ and

$$Z_2 = \begin{cases} [2i-1, 4i-2] \cup [n-4i+3, n-2i], & \text{for } i \in [2, (k+1)/2] \\ \{1,2\}, & \text{for } i = 1. \end{cases}$$

For the application of Step 3, $j' = \max\ s(W_2) = n - 4i + 3 > 2k$ and $j = \max\ s(Z_2) = 4i - 2$ in both cases. But $4i - 2 = 2k$ and $n - 4i + 3 = 2k + 1$ if $i = (k+1)/2$, and $4i - 2 < 2k$ if $i < (k+1)/2$. It follows that

$$I = \begin{cases} [0, k-1], & \text{if } i = (k+1)/2 \\ [0, 2i-1], & \text{if } i \in [2, k/2] \\ \{0\}, & \text{if } i = 1. \end{cases}$$

Thus

$$W_3 = \begin{cases} [n-2i+2, n-1] = [3k+1, n-1], & \text{if } i = (k+1)/2 \\ [n-2i+3, n-1], & \text{if } i \in [2, k/2] \\ \emptyset, & \text{if } i = 1 \end{cases}$$

and

$$Z_3 = \begin{cases} [2i-1, 4i-3] = [k, 2k-1], & \text{if } i = (k+1)/2 \\ [n-4i+3, n-2i], & \text{if } i \in [2, k/2] \\ \{1\}, & \text{if } i = 1. \end{cases}$$

(So the case $i = 1$ is finished.)

For the application of Step 4,

$$j' = \max\ s(W_3) = \begin{cases} 3k+1 > 2k, & \text{if } i = (k+1)/2 \\ n - 2i + 3 > 2k, & \text{if } i \in [2, k/2] \end{cases}$$

and

$$j = \max\ s(Z_3) = \begin{cases} 2k-1 < 2k, & \text{if } i = (k+1)/2 \\ n - 4i + 3 \geq 2k, & \text{if } i \in [2, k/2]. \end{cases}$$

It follows that

$$I = \begin{cases} [0, k-2], & \text{if } i = (k+1)/2 \\ [0, 2i-4], & \text{if } i \in [2, k/2]. \end{cases}$$

In all cases now $W_4 = \emptyset$ and we are finished.

Algorithm 6.11. This produces a proof in those cases excepted in Algorithm 6.8.

Step 1. This step is identical to Step 1 of Algorithm 6.8; that is, exploit inductive hypothesis (11, m) to its fullest extent to reduce the problem of showing $B_Z^m \geq C_W^m$ to showing $B_{Z_1}^m \geq C_{W_1}^m$, where $Z_1 = Z \setminus c(W)$ and $W_1 = W \setminus c(W)$.

Step 2. This step, in effect, exploits inductive hypothesis $(5', m)$(a), (f) and (g) applied to $j \in [2k+1, 3k]$ to its fullest extent. Formally, consider $B_{Z_1}^m - C_{W_1}^m$. If we set $P_1 = [2k+1, 3k] \cap Z_1 \cap W_1$, $P_1 \subseteq Z_1$, $P_1 \subseteq W_1$, and, by inductive hypothesis $(5', m)$(a), (f) and (g) we have $B_{P_1}^m \geq C_{P_1}^m$, thus

$$\begin{aligned} B_{Z_1}^m - C_{W_1}^m &= B_{Z_1}^m - B_{P_1}^m + B_{P_1}^m - C_{W_1}^m \\ &\geq B_{Z_1}^m - B_{P_1}^m + C_{P_1}^m - C_{W_1}^m \\ &= B_{Z_2}^m - C_{W_2}^m, \end{aligned}$$

where $Z_2 = Z_1 \setminus P_1$ and $W_2 = W_1 \setminus P_1$. To prove the result it now suffices to show $B^m_{Z_2} \geq C^m_{W_2}$.

Step 3. This step, in effect, exploits Claims (3, m+1) and (4, m+1) to their fullest extent. Formally, consider $B^m_{Z_2} - C^m_{W_2}$. Let $J_2 = \{j \in [1,k] : 2k+j \in Z_2 \text{ and } 2k-j \in W_2\}$, and let $P_2 = \{2k+j : j \in J_2\}$ and $Q_2 = \{2k-j : j \in J_2\}$. Note that $P_2 \subseteq Z_2$ and $Q_2 \subseteq W_2$. Note also that if $2k+j \in P_2$ and j is odd, $2k-j \in Q_2$. Thus, by Claim (3, m+1), $B^m_{\{2k+j\}} \geq C^m_{\{2k-j\}}$. Similarly if $2k+j \in P_2$ and j is even, $2k-j \in Q_2$ and $B^m_{\{2k+j\}} \geq C^m_{\{2k-j\}}$ by Claim (4, m+1). It follows that $B^m_{P_2} \geq C^m_{Q_2}$. Now

$$\begin{aligned} B^m_{Z_2} - C^m_{W_2} &= B^m_{Z_2} - B^m_{P_2} + B^m_{P_2} - C^m_{W_2} \\ &\geq B^m_{Z_2} - B^m_{P_2} + C^m_{Q_2} - C^m_{W_2} \\ &= B^m_{Z_3} - C^m_{W_3}, \end{aligned}$$

where $Z_3 = Z_2 \setminus P_2$ and $W_3 = W_2 \setminus Q_2$. To prove the result it now suffices to show $B^m_{Z_3} \geq C^m_{W_3}$.

It transpires that in all those cases excepted in Algorithm 6.8, application of Algorithm 6.11 to Step 3 either yields $Z_3 = W_3 = \emptyset$ and we are finished, or $Z_3 = \{j\}$ with $j \leq 2k$ and $W_3 = \{j-2\}$. In the latter case

$$\begin{aligned} C^m_{W_3} &= C^m_{\{j-2\}} \leq B^m_{\{j-1\}} \qquad \text{(by Corollary (6.5, m))} \\ &\leq B^m_{\{j\}} \qquad \text{(by Corollary 5.7)} \\ &= B^m_{Z_3}, \end{aligned}$$

and we are finished. Again we will illustrate with one of the cases.

EXAMPLE 6.12. Consider the excepted case in II(i) where $W = [2k-2i-1, 2k-2i] \cup [2k-2i+3, 2k+2i]$ and $Z = X_{2i} = [2k-2i+2, 2k+2i+1]$ for $i \in [1, (k-1)/2]$.

After Step 1 we found that

$$Z_1 = \begin{cases} Z \setminus [2k-2i+3, 2k+2i-3] & \\ \quad = \{2k-2i+2\} \cup [2k+2i-2, 2k+2i+1], & \text{for } i > 1 \\ Z = [2k, 2k+3], & \text{for } i = 1. \end{cases}$$

and

$$W_1 = \begin{cases} W \setminus [2k-2i+3, 2k+2i-3] & \\ \quad = [2k-2i-1, 2k-2i] \cup [2k+2i-2, 2k+2i], & \text{for } i > 1 \\ W = \{2k-3, 2k-2, 2k+1, 2k+2\}, & \text{for } i = 1. \end{cases}$$

After Step 2 we find

$$Z_2 = \begin{cases} Z_1 \setminus [2k+2i-2, 2k+2i] = \{2k-2i+2, 2k+2i+1\}, & \text{for } i > 1 \\ Z_1 \setminus \{2k+1, 2k+2\} = \{2k, 2k+3\}, & \text{for } i = 1 \end{cases}$$

and

$$W_2 = \begin{cases} W_1 \setminus [2k+2i-2, 2k+2i] = \{2k-2i-1, 2k-2i\}, & \text{for } i > 1 \\ W_1 \setminus \{2k+1, 2k+2\} = \{2k-3, 2k-2\}, & \text{for } i = 1. \end{cases}$$

After Step 3 we find $Z_3 = Z_2 \setminus \{2k+2i+1\} = \{2k-2i+2\}$ for $i \in [1, (k-1)/2]$ and $W_3 = W_2 \setminus \{2k-2i-1\} = \{2k-2i\}$ for $i \in [1, (k-1)/2]$, as required.

This completes the proof of Lemma (6.7, m). □

Proof of Claim (5′, m+1)(b), (c), (d), (e), (f), (g), (h) and (i).

(b) We wish to show that $\left|\left(B^{m+1}\right)^{(2i-1)}\right| \geq \left|\left(C^{m+1}\right)^{(2i-1)}\right|$ for $i \in [1, k/2]$, where $\langle a,b\rangle^{2i-1} \neq \langle x,y\rangle^{2i-1}$; that is, we wish to show $B^m_{X_{2i-1}} \geq C^m_{Y_{2i-1}}$ for $i \in [1, k/2]$, where $Y_{2i-1} \neq X_{2i-1}$.

Let $i \in [1, k/2]$. By Corollary 5.13 we can choose $W_{2i-1} \in T_{2i-1}$ such that

$$C^m_{W_{2i-1}} \geq C^m_{Y_{2i-1}}.$$

Further, if $W_{2i-1} = X_{2i-1}$, $B^m_{X_{2i-1}} \geq C^m_{X_{2i-1}} = C^m_{W_{2i-1}}$ by Claim (5′, m+1)(a), and if $W_{2i-1} \neq X_{2i-1}$, $W_{2i-1} \in T_{2i-1} \setminus \{X_{2i-1}\}$ and so

$$B^m_{X_{2i-1}} \geq C^m_{W_{2i-1}} \text{ by Lemma (6.7, m)(I)(i).}$$

Hence in all cases, $B^m_{X_{2i-1}} \geq C^m_{W_{2i-1}} \geq C^m_{Y_{2i-1}}$ as required.

(c) Is proved similarly to (b), using Corollary 5.13, Claim (5′, m+1)(a) and Lemma (6.7, m)(I)(ii).

(d) We wish to show that $\left|\left(B^{m+1}\right)^{(2i)}\right| \geq \left|\left(C^{m+1}\right)^{(2i)}\right|$ for $i \in [1, (k-1)/2]$, where $\langle a,b\rangle^{2i} \neq \langle x,y\rangle^{2i}$ and $\langle a,b\rangle^{2i} \neq \langle 2k-2i, 2k+2i-1\rangle$; that is, we wish to show $B^m_{X_{2i}} \geq C^m_{Y_{2i}}$ for $i \in [1, (k-1)/2]$, where $Y_{2i} \neq X_{2i}$ and $Y_{2i} \neq [2k-2i, 2k+2i-1]$.

Let $i \in [1, (k-1)/2]$. Since $Y_{2i} \neq [2k-2i, 2k+2i-1]$, by Corollary 5.13 we can choose $W_{2i} \in T_{2i}$ such that

$$C^m_{W_{2i}} \geq C^m_{Y_{2i}}.$$

Further, if $W_{2i} = X_{2i}$, $B^m_{X_{2i}} \geq C^m_{X_{2i}} = C^m_{W_{2i}}$ by Claim (5′, m+1)(a), and if $W_{2i} \neq X_{2i}$, $W_{2i} \in T_{2i} \setminus \{X_{2i}\}$ and so

$$B^m_{X_{2i}} \geq C^m_{W_{2i}} \text{ by Lemma (6.7, m)(I)(iii).}$$

Hence in all cases, $B^m_{X_{2i}} \geq C^m_{W_{2i}} \geq C^m_{Y_{2i}}$ as required.

(e) Is proved similarly to (d), using Corollary 5.13, Claim (5′, m+1)(a) and Lemma (6.7, m)(I)(iv).

(f) Is proved similarly to (b), using Corollary 5.13, Claim (5′, m+1)(a) and Lemma (6.7, m)(I)(v).

(g) Is proved similarly to (b), using Corollary 5.13, Claim (5′, m+1)(a) and Lemma (6.7, m)(I)(vi).

(h) We wish to show that $\left|\left(B^{m+1}\right)^{(2k-2i+1)}\right| \geq \left|\left(C^{m+1}\right)^{(2k-2i+1)}\right|$ for $i \in [1, (k+1)/2]$, where $\langle a,b\rangle^{2k-2i+1} \neq \langle x,y\rangle^{2k-2i+1}$; that is, we wish to show $B^m_{X_{2k-2i+1}} \geq C^m_{Y_{2k-2i+1}}$ for $i \in [1, (k+1)/2]$, where $Y_{2k-2i+1} \neq X_{2k-2i+1}$.

Let $i \in [1, (k+1)/2]$. Since $Y_{2k-2i+1} \neq X_{2k-2i+1}$, by Corollary 5.13 we can choose $W_{2k-2i+1} \in T'_{2k-2i+1}$ such that

$$C^m_{W_{2k-2i+1}} \geq C^m_{Y_{2k-2i+1}}.$$

Thus

$$B^m_{X_{2k-2i+1}} \geq C^m_{W_{2k-2i+1}} \geq C^m_{Y_{2k-2i+1}} \text{ by Lemma (6.7, m)(I)(vii).}$$

(i) Is proved similarly to (h), using Corollary 5.13 and Lemma (6.7, m)(I)(viii).

Proof of Claim (6, m+1). For $j \in [k, 2k-1]$ we have $\langle a,b\rangle^j = \langle x,y\rangle^j$ and $\langle a,b\rangle^{n-j} = \langle x,y\rangle^{n-j}$ by assumption.

(a) Let $i \in [1,(k+1)/2]$. Then

$$\begin{aligned}&\left|\left(B^{m+1}\right)^{(2k-2i+1)}\right| + \left|\left(B^{m+1}\right)^{(2k+2i-1)}\right| - \left|\left(C^{m+1}\right)^{(2k-2i+1)}\right| - \left|\left(C^{m+1}\right)^{(2k+2i-1)}\right| \\ &\quad = (B^m - C^m)[2i, n-2i+1] + (B^m - C^m)[2i-1, n-2i] \\ &\quad = (B^m - C^m)[2i-1, n-2i+1] + (B^m - C^m)[2i, n-2i] \\ &\quad \geq 0,\end{aligned}$$

by inductive hypothesis (11, m).

(b) Now let $i \in [1, k/2]$. Then

$$\begin{aligned}&\left|\left(B^{m+1}\right)^{(2k-2i)}\right| + \left|\left(B^{m+1}\right)^{(2k+2i)}\right| - \left|\left(C^{m+1}\right)^{(2k-2i)}\right| - \left|\left(C^{m+1}\right)^{(2k+2i)}\right| \\ &\quad = (B^m - C^m)[2i, n-2i-1] + (B^m - C^m)[2i+1, n-2i] \\ &\quad = (B^m - C^m)[2i, n-2i] + (B^m - C^m)[2i+1, n-2i-1] \\ &\quad \geq 0, \qquad \text{by inductive hypothesis (11, m).}\end{aligned}$$

Proof of Claim (7, m+1). Firstly note that under the given restrictions, in each case we wish to show

$$\left|\left(B^{m+1}\right)^{(j)}\right| + \left|\left(B^{m+1}\right)^{(s)}\right| - \left|\left(C^{m+1}\right)^{(j)}\right| - \left|\left(C^{m+1}\right)^{(s)}\right| \geq 0.$$

Now in each case

$$\begin{aligned}&\left|\left(B^{m+1}\right)^{(j)}\right| + \left|\left(B^{m+1}\right)^{(s)}\right| - \left|\left(C^{m+1}\right)^{(j)}\right| - \left|\left(C^{m+1}\right)^{(s)}\right| \\ &\quad = B^m_{X_j} + B^m_{X_s} - C^m_{Y_j} - C^m_{Y_s} \quad (\text{where } Y_s \neq X_s \text{ and } Y_j \text{ is as specified}) \\ &\quad \geq B^m_{X_j} + B^m_{X_s} - C^m_{Y_j} - C^m_{W},\end{aligned}$$

for some $W \in T'_s$ by (Corollary 5.13).

Thus it will suffice to show

$$B^m_{X_j} - C^m_{Y_j} + B^m_{X_s} - C^m_W \geq 0 \tag{6.1}$$

for all $W \in T'_s$, where Y_j is as specified. We call a proof of (6.1) for any given $W \in T'_s$ a **direct proof** (of a specific subcase of Claim (7, m+1)).

Note further that for $j = 2i$, $i \in [1,(k-1)/2]$, $Y_j = [2k-2i, 2k+2i-1]$ (as is the case throughout Claim (7, m+1) (a), (b) and (c)) and $W \in T'_s$

$$\begin{aligned}&B^m_{X_j} - C^m_{Y_j} + B^m_{X_s} - C^m_W \\ &\quad = B^m_{X_{2i}} - C^m_{[2k-2i,2k+2i-1]} + B^m_{X_s} - C^m_W \\ &\quad = B^m_{X_{2i}} - B^m_{X_{n-2i}} + \left(B^m_{X_{n-2i}} - C^m_{[2k-2i,2k+2i-1]}\right) + B^m_{X_s} - C^m_W \\ &\quad \geq B^m_{X_{2i}} - B^m_{X_{n-2i}} + B^m_{X_s} - C^m_W \quad (\text{by Lemma (6.7, m)(I)(ii)}) \\ &\quad = B^m_{[2k-2i+2,2k+2i+1]} - B^m_{[2k-2i+1,2k+2i]} + B^m_{X_s} - C^m_W \\ &\quad = B^m_{\{2k+2i+1\}} - B^m_{\{2k-2i+1\}} + B^m_{X_s} - C^m_W.\end{aligned}$$

Thus if we set $T'_s = (T'_s)_1 \cup (T'_s)_2$ for some choices of $(T'_s)_1$ and $(T'_s)_2$ (depending on s), and we can show (6.1) to be true for all $W \in (T'_s)_1$ and

$$B^m_{\{2k+2i+1\}} - B^m_{\{2k-2i+1\}} + B^m_{X_s} - C^m_W \geq 0 \tag{6.2}$$

for all $W \in (T'_s)_2$, (a proof of which, for some $W \in (T'_s)_2$, will be called an **indirect proof** of a specific subcase of Claim (7, m+1)(a), (b) or (c)), we will have completed the proof of Claim (7, m+1)(a), (b) and (c).

Similarly, throughout Claim (7, m+1) (d), (e) and (f) we have $j = n - 2i + 1$, $i \in [1, k/2]$ and $Y_j = [2k - 2i + 1, 2k + 2i - 2]$. So if $W \in T'_s$

$$\begin{aligned}
&B^m_{X_j} - C^m_{Y_j} + B^m_{X_s} - C^m_W \\
&\quad = B^m_{X_{n-2i+1}} - C^m_{[2k-2i+1,2k+2i-2]} + B^m_{X_s} - C^m_W \\
&\quad = B^m_{X_{n-2i+1}} - B^m_{X_{2i-1}} + \left(B^m_{X_{2i-1}} - C^m_{[2k-2i+1,2k+2i-2]}\right) + B^m_{X_s} - C^m_W \\
&\quad \geq B^m_{X_{n-2i+1}} - B^m_{X_{2i-1}} + B^m_{X_s} - C^m_W \quad \text{(by Lemma (6.7, m)(I)(i))} \\
&\quad = B^m_{[2k-2i+3,2k+2i]} - B^m_{[2k-2i+2,2k+2i-1]} + B^m_{X_s} - C^m_W \\
&\quad = B^m_{\{2k+2i\}} - B^m_{\{2k-2i+2\}} + B^m_{X_s} - C^m_W.
\end{aligned}$$

Thus again, if we can show (6.1) to be true for all $W \in (T'_s)_1$, and

$$B^m_{\{2k+2i\}} - B^m_{\{2k-2i+2\}} + B^m_{X_s} - C^m_W \geq 0 \tag{6.3}$$

for all $W \in (T'_s)_2$ (an indirect proof of the specific subcases of Claim (7, m+1)(d), (e) and (f)) we will have completed the proof of Claim (7, m+1)(d), (e) and (f).

Firstly we tackle the (easy) indirect cases.

(i) *Proof of Claim (7, m+1)(a) indirect cases.* Here $s = n - 2i$ and $i \in [1, (k-1)/2]$ so $X_s = X_{n-2i}$. We set $(T'_{n-2i})_1 = \emptyset$ and $(T'_{n-2i})_2 = T'_{n-2i}$ (So there are no direct cases to prove here). Let $W \in (T'_{n-2i})_2 = T'_{n-2i}$. Then

$$\begin{aligned}
&B^m_{\{2k+2i+1\}} - B^m_{\{2k-2i+1\}} + B^m_{X_s} - C^m_W \\
&\quad = B^m_{X_{2i}} - B^m_{X_{n-2i}} + B^m_{X_s} - C^m_W \\
&\quad = B^m_{X_{2i}} - C^m_W \quad (\text{since } X_s = X_{n-2i}) \\
&\quad \geq 0 \quad \text{(by Lemma (6.7, m)(II)(i))}.
\end{aligned}$$

(ii) *Proof of Claim (7, m+1)(d) indirect cases.* Here $s = 2i - 1$ and $i \in [1, k/2]$ so $X_s = X_{2i-1}$. We set $(T'_{2i-1})_1 = \emptyset$ and $(T'_{2i-1})_2 = T'_{2i-1}$ (So there are no direct cases to prove here). Let $W = (T'_{2i-1})_2 = T'_{2i-1}$. Then

$$\begin{aligned}
&B^m_{\{2k+2i\}} - B^m_{\{2k-2i+2\}} + B^m_{X_s} - C^m_W \\
&\quad = B^m_{X_{n-2i+1}} - B^m_{X_{2i-1}} + B^m_{X_s} - C^m_W \\
&\quad = B^m_{X_{n-2i+1}} - C^m_W \quad (\text{since } X_s = X_{2i-1}) \\
&\quad \geq 0 \quad \text{(by Lemma (6.7, m)(II)(ii))}.
\end{aligned}$$

(iii) *Proof of Claim (7, m+1)(c) indirect cases.* Here $s = 2k+2i$ and $i \in [1,(k-1)/2]$ so $X_s = X_{2k+2i} = [2i+1, n-2i]$. We set $(T'_{2k+2i})_1 = \{[2i-1, n-2i-2], [2i-1,2i]\cup[2i+3,n-2i], [1,n-1]\cap([2i+1,n-2i-2]\cup[n-2i+1,n-2i+2])\}$ and $(T'_{2k+2i})_2 = T'_{2k+2i}\setminus(T'_{2k+2i})_1$. Let $W \in (T'_{2k+2i})_2$. Then

$$\begin{aligned}
&B^m_{\{2k+2i+1\}} - B^m_{\{2k-2i+1\}} + B^m_{X_s} - C^m_W \\
&\quad = B^m_{\{2k+2i+1\}} - B^m_{\{2k-2i+1\}} + B^m_{[2i+1,n-2i]} - B^m_{[2i+2,n-2i+1]} \\
&\qquad + B^m_{[2i+2,n-2i+1]} - C^m_W \\
&\quad \geq B^m_{\{2k+2i+1\}} - B^m_{\{2k-2i+1\}} + B^m_{\{2i+1\}} - B^m_{\{n-2i+1\}} \\
&\qquad \text{(by Lemma (6.7, m)(II)(iii))} \\
&\quad = B^{m-1}_{[2i+1,n-2i-2]} - B^{m-1}_{[2i,n-2i+1]} + B^{m-1}_{[2k-2i,2k+2i+1]} - B^{m-1}_{[2k-2i+3,2k+2i]} \\
&\quad \text{(note that the formula is still valid in the case } 2i+1=k \\
&\quad \text{and } 2k+2i+1=3k) \\
&\quad = B^{m-1}_{\{2k-2i,2k-2i+1,2k-2i+2,2k+2i+1\}} - B^{m-1}_{\{2i,n-2i-1,n-2i,n-2i+1\}} \\
&\quad \geq 0
\end{aligned}$$

This follows by Lemma 6.2 since for $i \in [1,(k-1)/2]$, $\{2k-2i, 2k-2i+1, 2k-2i+2, 2k+2i+1\} \subseteq [k,3k]$ and $\{2i, n-2i-1, n-2i, n-2i+1\} \subseteq [1,k]\cup[3k,n-1]$.

(iv) *Proof of Claim (7, m+1)(b) indirect cases.* Here $s = 2k-2i$ and $i \in [1,(k-1)/2]$ so $X_s = X_{2k-2i} = [2i, n-2i-1]$. We set $(T'_{2k-2i})_1 = \{[2i+2, n-2i+1], [2i, n-2i-3]\cup[n-2i, n-2i+1]\}$ and $(T'_{2k-2i})_2 = T'_{2k-2i}\setminus(T'_{2k-2i})_1$. Let $W \in (T'_{2k-2i})_2$. Then

$$\begin{aligned}
&B^m{}_{\{2k+2i+1\}} - B^m_{\{2k-2i+1\}} + B^m_{X_s} - C^m_W \\
&\quad = -B^{m-1}_{\{2i,n-2i-1,n-2i,n-2i+1\}} + B^m_{[2i,n-2i-1]} - B^m_{[2i-1,n-2i-2]} \\
&\qquad + B^m_{[2i-1,n-2i-2]} - C^m_W \quad \text{(as in case (iii) above)} \\
&\quad \geq B^m_{\{n-2i-1\}} - B^m_{\{2i-1\}} - B^{m-1}_{\{2i,n-2i-1,n-2i,n-2i+1\}} \\
&\qquad \text{(by Lemma (6.7, m)(II)(iv))} \\
&\quad = \begin{cases} B^{m-1}_{[k,3k-1]} - B^{m-1}_{[k+3,3k-2]} - B^{m-1}_{\{k-1,3k,3k+1,3k+2\}}, & \text{if } i=(k-1)/2 \\ B^{m-1}_{[2k-2i+1,2k+2i+2]} - B^{m-1}_{[2k-2i+2,2k+2i-1]} & \\ \quad -B^{m-1}_{\{2i,n-2i-1,n-2i,n-2i+1\}}, & \text{if } i<(k-1)/2 \end{cases} \\
&\quad = \begin{cases} B^{m-1}_{\{k,k+1,k+2,3k-1\}} - B^{m-1}_{\{k-1,3k,3k+1,3k+2\}}, & \text{if } i=(k-1)/2 \\ B^{m-1}_{\{2k-2i+1\}\cup[2k+2i,2k+2i+2]} - B^{m-1}_{\{2i\}\cup[n-2i-1,n-2i+1]}, & \text{if } 2i\leq k-2 \end{cases} \\
&\quad \geq 0 \quad \text{(as in case (iii) above).}
\end{aligned}$$

(v) *Proof of Claim (7, m+1)(e) indirect cases.* Here $s = 2k-2i+1$ and $i \in [1,k/2]$ so $X_s = X_{2k-2i+1} = [2i, n-2i+1]$. We set

$$(T'_{2k-2i+1})_1 = \begin{cases} \{[2i-2, n-2i-1], & \\ \quad [2i-2,2i-1]\cup[2i+2,n-2i+1]\}, & \text{for } i>1 \\ \{[1,n-3], \{1\}\cup[4,n-1]\} = T'_{2k-1}, & \text{for } i=1 \end{cases}$$

and $(T'_{2k-2i+1})_2 = T'_{2k-2i+1} \setminus (T'_{2k-2i+1})_1$. Let $W \in (T'_{2k-2i+1})_2$ and note $(T'_{2k-2i+1})_2 = \emptyset$ if $i = 1$, so we may assume $i \in [2, k/2]$. Thus

$$\begin{aligned}
&B^m{}_{\{2k+2i\}} - B^m_{\{2k-2i+2\}} + B^m_{X_s} - C^m_W \\
&\quad = B^m_{\{2k+2i\}} - B^m_{\{2k-2i+2\}} + B^m_{[2i,n-2i+1]} - B^m_{[2i+1,n-2i+2]} \\
&\qquad + B^m_{[2i+1,n-2i+2]} - C^m_W \\
&\quad \geq B^m_{\{2k+2i\}} - B^m_{\{2k-2i+2\}} + B^m_{\{2i\}} - B^m_{\{n-2i+2\}} \quad \text{(by Lemma (6.7, m)(II)(v))} \\
&\quad = \begin{cases} B^{m-1}_{[k+1,3k]} - B^{m-1}_{[k-2,3k+1]} - B^{m-1}_{[k,3k-1]} - B^{m-1}_{[k+3,3k-2]}, & \text{if } i = k/2 \\ B^{m-1}_{[2i+1,n-2i]} - B^{m-1}_{[2i-2,n-2i+1]} + B^{m-1}_{[2k-2i+2,2k+2i+1]} & \\ -B^{m-1}_{[2k-2i+3,2k+2i-2]}, & \text{if } i < k/2 \end{cases} \\
&\quad = \begin{cases} B^{m-1}_{\{k,k+1,k+2,3k-1\}} - B^{m-1}_{\{k-2,k-1,k,3k+1\}}, & \text{if } i = k/2 \\ B^{m-1}_{\{2k-2i+2,2k+2i-1,2k+2i,2k+2i+1\}} - B^{m-1}_{\{2i-2,2i-1,2i,n-2i+1\}}, & \text{if } 2i \leq k-1 \end{cases} \\
&\quad \geq 0 \quad \text{(as in case (iii) above).}
\end{aligned}$$

(vi) *Proof of Claim (7, m+1)(f) indirect cases.* Here $s = 2k+2i-1$ and $i \in [1, k/2]$ so $X_s = X_{2k+2i-1} = [2i-1, n-2i]$. We set

$$(T'_{2k+2i-1})_1 = \begin{cases} \{[2i-1, n-2i-2] \cup [n-2i+1, n-2i+2], & \\ [2i+1, n-2i+2], [2i-3, 2i-2] \cup [2i+1, n-2i]\}, & \text{for } i > 1 \\ \{[3, n-1], [1, n-4] \cup \{n-1\}\} = T'_{2k+1}, & \text{for } i = 1. \end{cases}$$

and $(T'_{2k+2i-1})_2 = T'_{2k+2i-1} \setminus (T'_{2k+2i-1})_1$. Let $W \in (T'_{2k+2i-1})_2$ and note that $(T'_{2k+2i-1})_2 = \emptyset$ if $i = 1$, so we may assume $i \in [2, k/2]$.

$$\begin{aligned}
&B^m{}_{\{2k+2i\}} - B^m_{\{2k-2i+2\}} + B^m_{X_s} - C^m_W \\
&\quad = -B^{m-1}_{\{2i-2,2i-1,2i,n-2i+1\}} + B^m_{[2i-1,n-2i]} - B^m_{[2i-2,n-2i-1]} \\
&\qquad + B^m_{[2i-2,n-2i-1]} - C^m_W \text{(as in case (v) above)} \\
&\quad \geq B^m_{\{n-2i\}} - B^m_{\{2i-2\}} - B^{m-1}_{\{2i-2,2i-1,2i,n-2i+1\}} \quad \text{(by Lemma (6.7, m)(II)(vi))} \\
&\quad = B^{m-1}_{[2k-2i+1,2k+2i]} - B^{m-1}_{[2k-2i+4,2k+2i-1]} - B^{m-1}_{\{2i-2,2i-1,2i,n-2i+1\}} \\
&\quad = B^{m-1}_{\{2k-2i+1,2k-2i+2,2k-2i+3,2k+2i\}} - B^{m-1}_{\{2i-2,2i-1,2i,n-2i+1\}} \\
&\quad \geq 0 \quad \text{(as in case (iii) above).}
\end{aligned}$$

Finally it remains to prove Claim (7, m+1) in the complicated direct cases. As we have seen there is nothing to do for (7, m+1)(a) and (d). For (7, m+1)(b) and (c) we have $j = 2i$, $i \in [1, (k-1)/2]$ and $Y_j = [2k-2i, 2k+2i-1]$. Thus

$$\begin{aligned}
&B^m_{X_j} - C^m_{Y_j} + B^m_{X_s} - C^m_W \\
&\quad = B^m_{[2k-2i+2,2k+2i+1]} - C^m_{[2k-2i,2k+2i-1]} + B^m_{X_s} - C^m_W \\
&\quad \geq B^m_{\{2k+2i+1\}} - C^m_{\{2k-2i+1\}} + B^m_{X_s} - C^m_W \\
&\qquad \text{(applying the three steps of Algorithm 6.11 to the first two terms)} \\
&\quad \geq B^m_{\{2k+2i+1\}} - B^m_{\{2k+2i-1\}} + B^m_{X_s} - C^m_W \quad \text{(by Claim (3, m+1))} \\
&\quad = B^{m-1}_{[2i+1,n-2i-2]} - B^{m-1}_{[2i-1,n-2i]} + B^m_{X_s} - C^m_W \\
&\quad = B^m_{X_s} - C^m_W - B^{m-1}_{\{2i-1,2i,n-2i-1,n-2i\}} \\
&\quad \geq B^m_{X_s} - C^m_W - 4B^{m-1}_{\{k\}} \quad \text{(as in case (iii) above).}
\end{aligned}$$

It follows that to complete the proofs of (7, m+1)(b) and (c) it will suffice to show

$$B^m_{X_s} - C^m_W \geq 4B^{m-1}_{\{k\}}$$

for all $W \in (T'_s)_1$. Further for (7, m+1)(e) and (f) we have $j = n-2i+1$, $i \in [1, k/2]$ and $Y_j = [2k-2i+1, 2k+2i-2]$. Thus

$$\begin{aligned}
&B^m_{X_j} - C^m_{Y_j} + B^m_{X_s} - C^m_W \\
&\quad = B^m_{[2k-2i+3,2k+2i]} - C^m_{[2k-2i+1,2k+2i-2]} + B^m_{X_s} - C^m_W \\
&\quad \geq B^m_{\{2k+2i\}} - C^m_{\{2k-2i+2\}} + B^m_{X_s} - C^m_W \\
&\qquad \text{(applying the three steps of Algorithm 6.11 to the first two terms)} \\
&\quad \geq B^m_{\{2k+2i\}} - B^m_{\{2k+2i-2\}} + B^m_{X_s} - C^m_W \quad \text{(by Claim (4, m+1))} \\
&\quad = \begin{cases} B^{m-1}_{[2i+1,n-2i]} - B^{m-1}_{[2i-1,n-2i+2]} + B^m_{X_s} - C^m_W, & \text{if } i > 1 \\ B^{m-1}_{[3,n-2]} - B^{m-1}_{[1,n-1]} + B^m_{X_s} - C^m_W, & \text{if } i = 1 \end{cases} \\
&\quad = \begin{cases} B^m_{X_s} - C^m_W - B^{m-1}_{\{2i-1,2i,n-2i+1,n-2i+2\}}, & \text{if } i > 1 \\ B^m_{X_s} - C^m_W - B^{m-1}_{\{1,2,n-1\}}, & \text{if } i = 1 \end{cases} \\
&\quad \geq \begin{cases} B^m_{X_s} - C^m_W - 4B^{m-1}_{\{k\}}, & \text{if } i > 1 \text{ (as in (iii) above)} \\ B^m_{X_s} - C^m_W - 3B^{m-1}_{\{k\}}, & \text{if } i = 1. \end{cases}
\end{aligned}$$

Thus to complete the proof of (7, m+1) in all remaining cases it suffices to show

$$B^m_{X_s} - C^m_W \geq 4B^{m-1}_{\{k\}} \tag{6.4}$$

for all $W \in (T'_S)_1$.

Once more we provide an algorithm to achieve the desired result. The algorithm extends Algorithm 6.11 by also exploiting Claims (1, m+1) and (2, m+1) and the remainder of inductive hypothesis (5, m) to the full as follows:

ALGORITHM 6.13. Steps 1, 2, and 3 are the same as Algorithm 6.11, commencing with $Z = X_s$. So we are now considering $B^m_{Z_3} - C^m_{W_3}$, and wish to show $B^m_{Z_3} - C^m_{W_3} \geq 4B^{m-1}_{\{k\}}$.

Step 4. Consider $B^m_{Z_3} - C^m_{W_3}$. If $\{k\} \subseteq Z_3 \cap W_3$ then

$$\begin{aligned}
B^m_{Z_3} - C^m_{W_3} &= B^m_{Z_3} - B^m_{\{k\}} + B^m_{\{k\}} - C^m_{W_3} \\
&\geq B^m_{Z_3} - B^m_{\{k\}} + C^m_{\{k\}} - C^m_{W_3} \quad \text{(by inductive hypothesis (5, m))} \\
&= B^m_{Z_4} - C^m_{W_4}
\end{aligned}$$

where $Z_4 = Z_3 \setminus \{k\}$ and $W_4 = W_3 \setminus \{k\}$.

To prove the result it now suffices to show $B^m_{Z_4} - C^m_{W_4} \geq 4B^{m-1}_{\{k\}}$.

Step 5. Consider $B^m_{Z_4} - C^m_{W_4}$. If $k \in Z_4$ and $3k \in W_4$ then

$$\begin{aligned}
B^m_{Z_4} - C^m_{W_4} &= B^m_{Z_4} - B^m_{\{k\}} + B^m_{\{k\}} - C^m_{W_4} \\
&= B^m_{Z_4} - B^m_{\{k\}} + B^m_{\{3k\}} - C^m_{W_4} \quad \text{(by Lemma 6.2)} \\
&\geq B^m_{Z_4} - B^m_{\{k\}} + C^m_{\{3k\}} - C^m_{W_4} \quad \text{(by inductive hypothesis (5, m))} \\
&= B^m_{Z_5} - C^m_{W_5}
\end{aligned}$$

where $Z_5 = Z_4 \setminus \{k\}$ and $W_5 = W_4 \setminus \{3k\}$.

To prove the result it now suffices to show $B^m_{Z_5} - C^m_{W_5} \geq 4B^{m-1}_{\{k\}}$.

The next two steps now change in nature as we no longer concern ourselves about the nature of Z_5 or Z_6. In all cases to be considered it transpires that W_5 is a non-empty subset of $[1, k-1] \cup [3k+1, n-1]$.

Step 6. Consider $B^m_{Z_5} - C^m_{W_5}$. Let $P_5 = W_5 \cap (O[1, k-1] \cup E[3k+1, n-2])$. Now

$$\begin{aligned} B^m_{Z_5} - C^m_{W_5} &= B^m_{Z_5} - B^m_{P_5} + B^m_{P_5} - C^m_{W_5} \\ &\geq B^m_{Z_5} - B^m_{P_5} + C^m_{P_5} - C^m_{W_5} \quad \text{(by inductive hypothesis (5, m))} \\ &= B^m_{Z_6} - B^m_{P_6} - C^m_{W_6} \end{aligned}$$

where $Z_6 = Z_5 \setminus P_5$, $P_6 = P_5 \setminus Z_5$ and $W_6 = W_5 \setminus P_5$.

To prove the result it now suffices to show $B^m_{Z_6} - B^m_{P_6} - C^m_{W_6} \geq 4B^{m-1}_{\{k\}}$.

Step 7. Consider $B^m_{Z_6} - B^m_{P_6} - C^m_{W_6}$, $W_6 \subseteq E[2, k-1] \cup O[3k+1, n-1]$. Let $Q_6 = \{n - j : j \in W_6\}$. Now

$$\begin{aligned} B^m_{Z_6} - B^m_{P_6} - C^m_{W_6} &= B^m_{Z_6} - B^m_{P_6} - B^m_{Q_6} + B^m_{Q_6} - C^m_{W_6} \\ &\geq B^m_{Z_6} - B^m_{P_6} - B^m_{Q_6} \quad \text{(by Claims (1, m+1) and (2, m+1))} \\ &= B^m_{Z_7} - B^m_{P_6} - B^m_{Q_7} \end{aligned}$$

where $Z_7 = Z_6 \setminus Q_6$, and $Q_7 = Q_6 \setminus Z_6$.

To prove the result it now suffices to show $B^m_{Z_7} - B^m_{P_6} - B^m_{Q_7} \geq 4B^{m-1}_{\{k\}}$.

Step 8. Convert each term in $B^m_{Z_7}$, $B^m_{P_6}$, and $B^m_{Q_7}$, to level $m-1$ using $B^m_{\{j\}} = B^{m-1}_{X_j}$ and simplify. It will transpire that $B^m_{Z_7} - B^m_{P_7} - B^m_{Q_7}$ will simplify to a sum of terms of the form $B^{m-1}_{\{j\}}$. Furthermore, at least four of the $B^{m-1}_{\{j\}}$ terms to be summed will have $j \in [k, 3k]$, thus the overall result will follow using Lemma 6.2.

We will illustrate the use of Algorithm 6.13 in the cases (7, m+1), $W = [2i-1, n-2i-2] \cup [n-2i+1, n-2i+2]$ and $i > 1$, and $W = [3, n-1]$ with $i = 1$.

EXAMPLE 6.14. Proof of Claim (7, m+1)(f) in the case $W = [2i-1, n-2i-2] \cup [n-2i+1, n-2i+2]$ and $i > 1$. Here $s = 2k+2i-1$ and $i \in [1, k/2]$ so $X_s = X_{2k+2i-1} = [2i-1, n-2i]$ and $i \in [2, k/2]$. Thus

$$B^m_Z - C^m_W = B^m_{X_s} - C^m_W = B^m_{[2i-1,n-2i]} - C^m_{[2i-1,n-2i-2]\cup[n-2i+1,n-2i+2]}.$$

After Step 1 we found that $Z_1 = Z \setminus [2i+2, n-2i-2] = [2i-1, 2i+1] \cup [n-2i-1, n-2i]$ and $W_1 = W \setminus [2i+2, n-2i-2] = [2i-1, 2i+1] \cup [n-2i+1, n-2i+2]$.

After Step 2 we find $Z_2 = Z_1$, $W_2 = W_1$ (since $[2k+1, 3k] \cap W_1 = \emptyset$).

Applying Step 3 we find

$$J_2 = \begin{cases} \emptyset, & \text{if } 2i+1 < k \\ \{k\}, & \text{if } 2i+1 = k \\ \{k-1, k\}, & \text{if } 2i = k. \end{cases}$$

So

$$Z_3 = \begin{cases} Z_2, & \text{if } 2i+1 < k \\ Z_2 \setminus \{3k\} = [k-2, k] \cup \{3k+1\}, & \text{if } 2i+1 = k \\ Z_2 \setminus \{3k-1, 3k\} = [k-1, k+1], & \text{if } 2i = k \end{cases}$$

and

$$W_3 = \begin{cases} W_2, & \text{if } 2i+1 < k \\ W_2 \setminus \{k\} = [k-2, k-1] \cup [3k+2, 3k+3], & \text{if } 2i+1 = k \\ W_2 \setminus \{k, k+1\} = \{k-1, 3k+1, 3k+2\}, & \text{if } 2i = k. \end{cases}$$

In all cases $W_3 \subseteq [1, k-1] \cup [3k+1, n-1]$ so Steps 4 and 5 leave W and Z unchanged; that is $W_5 = W_3$ and $Z_5 = Z_3$. Applying Step 6 we find

$$P_5 = \begin{cases} \{2i-1, 2i+1, n-2i+2\}, & \text{if } 2i+1 < k \\ \{k-2, 3k+3\}, & \text{if } 2i+1 = k \\ \{k-1, 3k+2\}, & \text{if } 2i = k. \end{cases}$$

Therefore

$$Z_6 = Z_5 \setminus P_5 = \begin{cases} \{2i, n-2i-1, n-2i\}, & \text{if } 2i+1 < k \\ \{k-1, k, 3k+1\}, & \text{if } 2i+1 = k \\ \{k, k+1\}, & \text{if } 2i = k, \end{cases}$$

$$P_6 = P_5 \setminus Z_5 = \begin{cases} \{n-2i+2\}, & \text{if } 2i+1 < k \\ \{3k+3\}, & \text{if } 2i+1 = k \\ \{3k+2\}, & \text{if } 2i = k \end{cases}$$

and

$$W_6 = W_5 \setminus P_5 = \begin{cases} \{2i, n-2i+1\}, & \text{if } 2i+1 < k \\ \{k-1, 3k+2\}, & \text{if } 2i+1 = k \\ \{3k+1\}, & \text{if } 2i = k. \end{cases}$$

Applying Step 7 we find

$$Q_6 = \begin{cases} \{2i-1, n-2i\}, & \text{if } 2i+1 < k \\ \{k-2, 3k+1\}, & \text{if } 2i+1 = k \\ \{k-1\}, & \text{if } 2i = k. \end{cases}$$

So

$$Z_7 = Z_6 \setminus Q_6 = \begin{cases} \{2i, n-2i-1\}, & \text{if } 2i+1 < k \\ \{k-1, k\}, & \text{if } 2i+1 = k \\ \{k, k+1\}, & \text{if } 2i = k \end{cases}$$

and

$$Q_7 = Q_6 \setminus Z_6 = \begin{cases} \{2i-1\}, & \text{if } 2i+1 < k \\ \{k-2\}, & \text{if } 2i+1 = k \\ \{k-1\}, & \text{if } 2i = k; \end{cases}$$

that is,

$$B^m_{Z_7} - B^m_{P_6} - B^m_{Q_7} = \begin{cases} B^m_{\{2i, n-2i-1\}} - B^m_{\{2i-1, n-2i+2\}}, & \text{if } 2i+1 < k \\ B^m_{\{k-1, k\}} - B^m_{\{k-2, 3k+3\}}, & \text{if } 2i+1 = k \\ B^m_{\{k, k+1\}} - B^m_{\{k-1, 3k+2\}}, & \text{if } 2i = k \end{cases}$$

and we wish to show $B^m_{Z_7} - B^m_{P_6} - B^m_{Q_7} \geq 4B^{m-1}_{\{k\}}$.

Applying Step 8 gives

$$B^m_{Z_7} - B^m_{P_6} - B^m_{Q_7} = \begin{cases} B^{m-1}_{X_{2i}} + B^{m-1}_{X_{n-2i-1}} - B^{m-1}_{X_{2i-1}} - B^{m-1}_{X_{n-2i+2}}, & \text{if } 2i+1 < k \\ B^{m-1}_{X_{k-1}} + B^{m-1}_{X_k} - B^{m-1}_{X_{k-2}} - B^{m-1}_{X_{3k+3}}, & \text{if } 2i+1 = k \\ B^{m-1}_{X_k} + B^{m-1}_{X_{k+1}} - B^{m-1}_{X_{k-1}} - B^{m-1}_{X_{3k+2}}, & \text{if } 2i = k \end{cases}$$

$$= \begin{cases} B^{m-1}_{[2k-2i+2,2k+2i+1]} + B^{m-1}_{[2k-2i+1,2k+2i+2]} - B^{m-1}_{X_{[2k-2i+2,2k+2i-1]}} & \\ \quad -B^{m-1}_{X_{[2k-2i+3,2k+2i-2]}}, & \text{if } 2i+1 < k \\ B^{m-1}_{[k+3,3k]} + B^{m-1}_{[k+1,3k]} - B^{m-1}_{[k+3,3k-2]} - B^{m-1}_{[k+4,3k-3]}, & \text{if } 2i+1 = k \\ B^{m-1}_{[k,3k-1]} + B^{m-1}_{[k,3k+1]} - B^{m-1}_{[k+2,3k-1]} - B^{m-1}_{[k+3,3k-2]}, & \text{if } 2i = k \end{cases}$$

$$= \begin{cases} B^{m-1}_{[2k+2i,2k+2i+1]} + B^{m-1}_{[2k-2i+1,2k-2i+2]\cup[2k+2i-1,2k+2i+2]}, & \text{if } 2i+2 \leq k \\ B^{m-1}_{[3k-1,3k]} + B^{m-1}_{\{k+1,k+2,k+3,3k-2,3k-1,3k\}}, & \text{if } 2i+1 = k \\ B^{m-1}_{[k,k+1]} + B^{m-1}_{\{k,k+1,k+2,3k-1,3k,3k+1\}}, & \text{if } 2i = k \end{cases}$$

$$\geq 7B^{m-1}_{\{k\}} \quad \text{(by Lemma 6.2)}.$$

EXAMPLE 6.15. Proof of Claim (7,m+1)(f) in the case $W = [3, n-1]$ and $i = 1$. Here $s = 2k+1$ so $X_s = X_{2k+1} = [1, n-2]$. Therefore

$$B^m_Z - C^m_W = B^m_{X_s} - C^m_W = B^m_{[1,n-2]} - C^m_{[3,n-1]}.$$

After Step 1 we found that $Z_1 = Z \setminus [3, n-3] = \{1, 2, n-2\}$ and $W_1 = W \setminus [3, n-3] = \{n-2, n-1\}$.

After Step 2 we find

$$Z_2 = \begin{cases} Z_1 = \{1, 2, n-2\}, & \text{if } k > 2 \\ Z_1 \setminus \{n-2\} = \{1, 2\}, & \text{if } k = 2 \end{cases}$$

and

$$W_2 = \begin{cases} W_1 = \{n-2, n-1\}, & \text{if } k > 2 \\ W_1 \setminus \{n-2\} = \{n-1\}(= \{7\}), & \text{if } k = 2. \end{cases}$$

In all cases $W_2 \subseteq [1, k-1] \cup [3k+1, n-1]$, so Steps 3, 4 and 5 leave W and Z unchanged; that is, $W_5 = W_2$ and $Z_5 = Z_2$.

Applying Step 6 we find

$$P_5 = \begin{cases} \{n-2\}, & \text{if } k > 2 \\ \emptyset, & \text{if } k = 2. \end{cases}$$

Thus $Z_6 = Z_5 \setminus P_5 = \{1, 2\}$ in all cases, $P_6 = P_5 \setminus Z_5 = \emptyset$ in all cases and $W_6 = W_5 \setminus P_5 = \{n-1\}$ in all cases.

Applying Step 7 we find $Q_6 = \{1\}$, $Z_7 = Z_6 \setminus Q_6 = \{2\}$ and $Q_7 = Q_6 \setminus Z_6 = \emptyset$. Thus

$$B^m_{Z_7} - B^m_{P_6} - B^m_{Q_7} = B^m_{Z_7} = B^m_{\{2\}}$$

in all cases, so we wish to show $B^m_{\{2\}} \geq 4B^{m-1}_{\{k\}}$.

Applying Step 8,

$$B^m_{\{2\}} = \begin{cases} B^{m-1}_{[2k,2k+3]}, & \text{if } k > 2 \\ B^{m-1}_{[2,5]}, & \text{if } k = 2 \end{cases}$$
$$\geq 4B^{m-1}_{\{k\}} \quad \text{(by Lemma 6.2).}$$

Proof of Claim (8, m+1). By Claim (5, m+1)(c) and (d), $\left|\left(B^{m+1}\right)^{(2k-j)}\right| \geq \left|\left(C^{m+1}\right)^{(2k-j)}\right|$ for $j \in [1, k-1]$ unless both

$$Y_{2k-j} = X_{2k-j} \text{ and } Y_{2k+j} = X_{2k+j}.$$

Also, by Claim (5, m+1) $\left|\left(B^{m+1}\right)^{(2k+j)}\right| \geq \left|\left(C^{m+1}\right)^{(2k+j)}\right|$ for $j \in [1, k-1]$. Further, by Claim (6, m+1)(a) and (b)

$$\left|\left(B^{m+1}\right)^{(2k-j)}\right| + \left|\left(B^{m+1}\right)^{(2k+j)}\right| \geq \left|\left(C^{m+1}\right)^{(2k-j)}\right| + \left|\left(C^{m+1}\right)^{(2k+j)}\right|$$

for $j \in [1, k-1]$ provided both $Y_{2k-j} = X_{2k-j}$ and $Y_{2k+j} = X_{2k+j}$.

Combining these points we deduce

$$\left|\left(B^{m+1}\right)^{(2k-j)}\right| + \left|\left(B^{m+1}\right)^{(2k+j)}\right| \geq \left|\left(C^{m+1}\right)^{(2k-j)}\right| + \left|\left(C^{m+1}\right)^{(2k+j)}\right| \tag{6.5}$$

for $j \in [1, k-1]$.

Further, by Claim (5,m+1),

$$\left|\left(B^{m+1}\right)^{(k)}\right| \geq \left|\left(C^{m+1}\right)^{(k)}\right|, \qquad \left|\left(B^{m+1}\right)^{(2k)}\right| \geq \left|\left(C^{m+1}\right)^{(2k)}\right|$$

and

$$\left|\left(B^{m+1}\right)^{(3k)}\right| \geq \left|\left(C^{m+1}\right)^{(3k)}\right|$$

.

Simple appropriate summation involving these latter four inequalities now establishes Claim (8, m+1).

Proof of Claim (9, m+1). Note that by symmetry it suffices to prove Claim (9, m+1) for $j \in E[2, k-1] \cup O[3k+1, n-1]$.

Case 1. For $j \in E[2, k-1]$, (that is, $j = 2i$ for $i \in [1, (k-1)/2]$).

Note that $\left|\left(B^{m+1}\right)^{(n-j)}\right| \geq \left|\left(C^{m+1}\right)^{(n-j)}\right|$ by Claim (5, m+1).

(a) If $Y_{2i} \neq [2k-2i, 2k+2i-1]$ then $\left|\left(B^{m+1}\right)^{(j)}\right| \geq \left|\left(C^{m+1}\right)^{(j)}\right|$ by Claim (5, m+1)(a). Combining this with the above and with (6.5) establishes Claim (9, m+1).

(b) If $Y_{2i} = [2k-2i, 2k+2i-1]$, then by Corollary 4.13, for at least one $s \in \{2k-2i, 2k+2i, n-2i\}$, $Y_s \neq X_s$, and so, by Claim (7, m+1), $\left|\left(B^{m+1}\right)^{(j)}\right| + \left|\left(B^{m+1}\right)^{(s)}\right| \geq \left|\left(C^{m+1}\right)^{(j)}\right| + \left|\left(C^{m+1}\right)^{(s)}\right|$.

For $s = n - 2i$, the above inequality and (6.5) establishes Claim (9, m+1).

For $s \in \{2k-2i, 2k+2i\}$, $Y_s \neq X_s$ gives both $\left|\left(B^{m+1}\right)^{(2k-2i)}\right| \geq \left|\left(C^{m+1}\right)^{(2k-2i)}\right|$ and $\left|\left(B^{m+1}\right)^{(2k+2i)}\right| \geq \left|\left(C^{m+1}\right)^{(2k+2i)}\right|$, so the above inequality, one of these inequalities and the fact that $\left|\left(B^{m+1}\right)^{(n-2i)}\right| \geq \left|\left(C^{m+1}\right)^{(n-2i)}\right|$ establish Claim (9, m+1).

Case 2. For $j \in O[3k+1, n-1]$, (that is, $j = n - 2i + 1$ for $i \in [1, k/2]$). Note that $\left|\left(B^{m+1}\right)^{(n-j)}\right| \geq \left|\left(C^{m+1}\right)^{(n-j)}\right|$ by Claim (5, m+1).

(a) If $Y_{n-2i+1} \neq [2k-2i+1, 2k+2i-2]$, $\left|\left(B^{m+1}\right)^{(j)}\right| \geq \left|\left(C^{m+1}\right)^{(j)}\right|$ by Claim (5, m+1)(b). Combining this with the above and with (6.5) establishes Claim (9, m+1).

(b) If $Y_{n-2i+1} = [2k-2i+1, 2k+2i-2]$, then by Corollary 4.13, for at least one $s \in \{2k-2i+1, 2k+2i-1, 2i-1\}$, $Y_s \neq X_s$, and so, by Claim (7, m+1), $\left|\left(B^{m+1}\right)^{(j)}\right| + \left|\left(B^{m+1}\right)^{(s)}\right| \geq \left|\left(C^{m+1}\right)^{(j)}\right| + \left|\left(C^{m+1}\right)^{(s)}\right|$.

For $s = 2i - 1$, the above inequality and (6.5) establishes Claim (9, m+1).

For $s = 2k - 2i + 1$ or $2k + 2i - 1$, $Y_s \neq X_s$ gives both $\left|\left(B^{m+1}\right)^{(2k-2i+1)}\right| \geq \left|\left(C^{m+1}\right)^{(2k-2i+1)}\right|$ and $\left|\left(B^{m+1}\right)^{(2k+2i-1)}\right| \geq \left|\left(C^{m+1}\right)^{(2k+2i-1)}\right|$, so the above inequality together with one of these inequalities and the fact that $\left|\left(B^{m+1}\right)^{(2i-1)}\right| \geq \left|\left(C^{m+1}\right)^{(2i-1)}\right|$ establish Claim (9, m+1).

Proof of Claim (10, m+1). This follows directly from the fact that

$$\left|\left(B^{m+1}\right)^{(k)}\right| + \left|\left(B^{m+1}\right)^{(3k)}\right| \geq \left|\left(C^{m+1}\right)^{(k)}\right| + \left|\left(C^{m+1}\right)^{(3k)}\right|$$

and Claim (9, m+1) using summation.

Proof of Claim (11, m+1). This follows directly from Claim (8, m+1) and Claim (10, m+1) using summation.

This completes the proof of Lemma 6.4. □

7. The remaining case

A simple corollary of Lemma 6.4 (xi) in the case $j = 1$, Notation 5.1, Lemma 5.4 and the comments following it, is that for $n \equiv 0 \pmod 4$ the permutation θ_n has maximum entropy amongst those n-cycles ϕ which are maximodal and for which $\phi(1) < \phi(2)$. If we can show that the entropy of θ_n is at least as great as the entropy of any n-cycle ϕ which is maximodal and for which $\phi(1) > \phi(2)$, then by Theorem 2.10 and the paragraph immediately above it, θ_n will have maximum entropy among all n-cycles ϕ. Theorem 4.2 will then be complete provided we can show $\widetilde{\theta_n}$, $\overline{\theta_n}$ and $\theta_n^* = \widetilde{\overline{\theta_n}}$ are all cycles with the same entropy as θ_n. It turns out that θ_n^* has a central role in the completion of our task.

LEMMA 7.1. *For $i \in [1, n]$, $\theta^*(i) = n + 1 - \theta(i)$.*

LEMMA 7.2. *For a permutation $\theta \in P_n$, n even, $(\theta^*)^* = \theta$ and θ^* is maximodal with $\theta^*(1) > \theta^*(2)$ if and only if θ is maximodal with $\theta(1) < \theta(2)$.*

LEMMA 7.3. *Let $\theta \in P_n$ and let C be the induced matrix of θ and D the induced matrix of θ^*. Then for $i, j \in [1, n-1]$*

$$c_{i\,j} = 1 \iff d_{i\,n-j} = 1.$$

NOTATION 7.4. If C is an $(n-1) \times (n-1)$ matrix whose only entries are 0 and 1 then C^* is that $(n-1) \times (n-1)$ matrix given by

$$c^*_{i\,j} = 1 \iff c_{i\,n-j} = 1,$$

for all $i, j \in [1, n-1]$.

LEMMA 7.5. *If C is an $(n-1) \times (n-1)$ matrix whose only entries are 0 and 1 then for all $p \in \mathbb{N} \cup \{0\}$ and all $j \in [1, n-1]$*

$$\left|(C^{*p})^{(j)}\right| = \left|(C^p)^{(n-j)}\right|.$$

COROLLARY 7.6. *If C is an $(n-1) \times (n-1)$ matrix whose only entries are 0 and 1 then for all $p \in \mathbb{N} \cup \{0\}$*

$$\|C^{*p}\| = \|C^p\|.$$

COROLLARY 7.7. *If B^* is the induced matrix of θ_n^*, then for all $p \in \mathbb{N} \cup \{0\}$,*

$$\|B^{*p}\| = \|B^p\|$$

and hence θ_n and θ_n^* have the same entropy.

NOTATION 7.8. If Γ is the class of $(n-1) \times (n-1)$ matrices used in Lemma 6.4 then

$$\Gamma^* = \{C^* : C \in \Gamma\}.$$

COROLLARY 7.9. *For all $C^* \in \Gamma^*$ and all $p \in \mathbb{N} \cup \{0\}$*

$$\|C^{*p}\| = \|C^p\| \leq \|B^p\| = \|B^{*p}\|.$$

LEMMA 7.10. *Let D be the induced matrix of an n-permutation ϕ such that ϕ is maximodal with $\phi(1) > \phi(2)$ and*

(i) *For some $i \in [1, (k-1)/2]$ the matrix D is identical to the matrix A^* on each of the four columns $2i, 2k-2i, 2k+2i, n-2i$*

or

(ii) *For some $i \in [1, k/2]$ the matrix D is identical to the matrix A^* on each of the four columns $2i-1, 2k-2i+1, 2k+2i-1, n-2i+1$.*

Then ϕ is not a cycle.

PROOF. First note that by Lemma 7.3, the matrix D is identical to the matrix A^* on column j if and only if the matrix D^* is identical to the matrix A on column $n-j$, so (i) and (ii) above are equivalent to $(i)'$ and $(ii)'$ where $(i)'$ and $(ii)'$ are obtained from (i) and (ii) by replacing D with D^* and A^* with A throughout. A consequence of this is that ϕ^* satisfies precisely the conditions of Proposition 4.12, which establishes ϕ^* is not a cycle since for some $i \in [1,(k-1)/2]$, $[2i+1, 2k+2i] \cup [2k+2i+1, n-2i]$ is fully invariant under ϕ^*, or for some $i \in [1, k/2]$, $[2i, 2k-2i+1] \cup [2k+2i, n-2i+1]$ is fully invariant under ϕ^*. However $\phi = \rho \circ \phi^*$ where $\rho(j) = n+1-j$ for all $j \in [1, n]$ and the above mentioned sets are trivially fully invariant under ρ. Thus they are also fully invariant under ϕ and hence ϕ is not a cycle. □

COROLLARY 7.11. *The set of all maximodal n-cycles ϕ for which $\phi(1) > \phi(2)$ is a subset of the set of all maximodal n-permutations θ for which $\theta(1) > \theta(2)$ and for which their induced matrices $M(\theta)$ satisfy the conditions*

(i) *For all $i \in [1,(k-1)/2]$ at least one of the following inequalities holds:*

$$M(\theta)^{(2i)} \neq A^{*(2i)}$$
$$M(\theta)^{(2k-2i)} \neq A^{*(2k-2i)}$$
$$M(\theta)^{(2k+2i)} \neq A^{*(2k+2i)}$$
$$M(\theta)^{(n-2i)} \neq A^{*(n-2i)}$$

and

(ii) *For all $i \in [1, k/2]$ at least one of the following inequalities holds:*

$$M(\theta)^{(2i-1)} \neq A^{*(2i-1)}$$
$$M(\theta)^{(2k-2i+1)} \neq A^{*(2k-2i+1)}$$
$$M(\theta)^{(2k+2i-1)} \neq A^{*(2k+2i-1)}$$
$$M(\theta)^{(n-2i+1)} \neq A^{*(n-2i+1)}.$$

LEMMA 7.12. *Let θ be a maximodal n-permutation for which $\theta(1) > \theta(2)$ and let $M(\theta)$ be its induced matrix. Let $M(\theta)$ satisfy conditions (i) and (ii) of Corollary 7.11. Then there exists an element $C^* \in \Gamma^*$ such that C^* dominates $M(\theta)$.*

PROOF. The n-permutation θ^* is maximodal with $\theta^*(1) < \theta^*(2)$ and its induced matrix $M(\theta^*)$ satisfies condition 5 in the definition of Γ. Thus by the evident and valid strengthened version of Lemma 5.4, there is an element $C \in \Gamma$ such that C dominates $M(\theta^*)$. But now $C^* \in \Gamma^*$ and by Corollary 7.6, C^* dominates $M(\theta)$ as required. □

A combination of Corollaries 7.9 and 7.11 and Lemma 7.12 now completes our task of showing θ_n has maximum entropy among all n-cycles.

We have already seen the entropy of θ_n and the entropy of θ_n^* are identical. Combining this with the fact that taking duals is an entropy preserving idempotent operation shows $\overline{\theta_n}$ and $\widetilde{\theta_n}$ also have the same entropy. Finally, to show $\widetilde{\theta_n}$, $\overline{\theta_n}$ and θ_n^* are all cycles, note that $\theta_n(i) = j \iff \overline{\theta_n}(n+1-i) = n+1-j$ (trivially the reason that θ is a cycle if and only if $\overline{\theta}$ is a cycle), $\iff \theta_n^*(i) = n+1-j \iff \widetilde{\theta_n}(n+1-i) = j$, and that θ_n is a cycle implies $\overline{\theta_n}$ is a cycle, $\widetilde{\theta_n}$ is a cycle implies θ_n^* is a cycle.

References

[1] S. Baldwin. Generalizations of a theorem of Sarkovskii on orbits of continuous real-valued functions. *Discrete Math*, **67**, (1987), 111-127.

[2] L. Block and A. Coppel. Dynamics in One Dimension. *Lecture Notes in Math.*, **1513**, Springer-Verlag, Berlin and New York, (1992).

[3] L. Block, J. Guckenheimer, M. Misiurewicz and L.S. Young. Periodic points and topological entropy for one-dimensional maps. *Lecture Notes in Math.*, **819**, Springer-Verlag, Berlin and New York, (1980), 18-34.

[4] W. Geller and J. Tolosa. Maximal Entropy Odd Orbit Types. *Transactions Amer. Math. Soc.*, **329**, No.1, (1992), 161-171.

[5] W. Geller and B. Weiss. Uniqueness of maximal entropy odd orbit types. *Proc. Amer. Math. Soc.*, **123**, No. 6, (1995), 1917-1922.

[6] W. Geller and Z. Zhang. Maximal entropy permutations of even size. *Proc. Amer. Math. Soc.*, details

[7] I. Jungreis. Some Results on the Sarkovskii Partial Ordering of Permutations. *Transactions Amer. Math. Soc.*, **325**, No.1, (1991), 319-344.

[8] D. M. King. Maximal Entropy of permutations of even order. *Ergod. Th. & Dynam. Sys.*, **17**, No.6, (1997), 1409-1417.

[9] D. M. King. Non-uniqueness of even order permutations with maximal entropy. *Ergod. Th. & Dynam. Sys.*, (to appear).

[10] M. Misiurewicz and Z. Nitecki. Combinatorial Patterns for maps of the Interval. *Memoirs Amer. Math. Soc.*, **94**, No.456, (1991).

[11] M. Misiurewicz and W. Szlenk. Entropy of piecewise monotone mappings. *Studia Math.*, **67**, (1980), 45-63.

Editorial Information

To be published in the *Memoirs*, a paper must be correct, new, nontrivial, and significant. Further, it must be well written and of interest to a substantial number of mathematicians. Piecemeal results, such as an inconclusive step toward an unproved major theorem or a minor variation on a known result, are in general not acceptable for publication. Papers appearing in *Memoirs* are generally longer than those appearing in *Transactions*, which shares the same editorial committee.

As of March 31, 2001, the backlog for this journal was approximately 6 volumes. This estimate is the result of dividing the number of manuscripts for this journal in the Providence office that have not yet gone to the printer on the above date by the average number of monographs per volume over the previous twelve months, reduced by the number of volumes published in four months (the time necessary for preparing a volume for the printer). (There are 6 volumes per year, each containing at least 4 numbers.)

A Consent to Publish and Copyright Agreement is required before a paper will be published in the *Memoirs*. After a paper is accepted for publication, the Providence office will send a Consent to Publish and Copyright Agreement to all authors of the paper. By submitting a paper to the *Memoirs*, authors certify that the results have not been submitted to nor are they under consideration for publication by another journal, conference proceedings, or similar publication.

Information for Authors

Memoirs are printed from camera copy fully prepared by the author. This means that the finished book will look exactly like the copy submitted.

The paper must contain a *descriptive title* and an *abstract* that summarizes the article in language suitable for workers in the general field (algebra, analysis, etc.). The *descriptive title* should be short, but informative; useless or vague phrases such as "some remarks about" or "concerning" should be avoided. The *abstract* should be at least one complete sentence, and at most 300 words. Included with the footnotes to the paper should be the 2000 *Mathematics Subject Classification* representing the primary and secondary subjects of the article. The classifications are accessible from `www.ams.org/msc/`. The list of classifications is also available in print starting with the 1999 annual index of *Mathematical Reviews*. The Mathematics Subject Classification footnote may be followed by a list of *key words and phrases* describing the subject matter of the article and taken from it. Journal abbreviations used in bibliographies are listed in the latest *Mathematical Reviews* annual index. The series abbreviations are also accessible from `www.ams.org/publications/`. To help in preparing and verifying references, the AMS offers MR Lookup, a Reference Tool for Linking, at `www.ams.org/mrlookup/`. When the manuscript is submitted, authors should supply the editor with electronic addresses if available. These will be printed after the postal address at the end of the article.

Electronically prepared manuscripts. The AMS encourages electronically prepared manuscripts, with a strong preference for $\mathcal{A}_{\mathcal{M}}\mathcal{S}$-LaTeX. To this end, the Society has prepared $\mathcal{A}_{\mathcal{M}}\mathcal{S}$-LaTeX author packages for each AMS publication. Author packages include instructions for preparing electronic manuscripts, the *AMS Author Handbook*, samples, and a style file that generates the particular design specifications of that publication series. Though $\mathcal{A}_{\mathcal{M}}\mathcal{S}$-LaTeX is the highly preferred format of TeX, author packages are also available in $\mathcal{A}_{\mathcal{M}}\mathcal{S}$-TeX.

Authors may retrieve an author package from e-MATH starting from www.ams.org/tex/ or via FTP to ftp.ams.org (login as anonymous, enter username as password, and type cd pub/author-info). The *AMS Author Handbook* and the *Instruction Manual* are available in PDF format following the author packages link from www.ams.org/tex/. The author package can be obtained free of charge by sending email to pub@ams.org (Internet) or from the Publication Division, American Mathematical Society, P.O. Box 6248, Providence, RI 02940-6248. When requesting an author package, please specify AMS-LaTeX or AMS-TeX, Macintosh or IBM (3.5) format, and the publication in which your paper will appear. Please be sure to include your complete mailing address.

Sending electronic files. After acceptance, the source file(s) should be sent to the Providence office (this includes any TeX source file, any graphics files, and the DVI or PostScript file).

Before sending the source file, be sure you have proofread your paper carefully. The files you send must be the EXACT files used to generate the proof copy that was accepted for publication. For all publications, authors are required to send a printed copy of their paper, which exactly matches the copy approved for publication, along with any graphics that will appear in the paper.

TeX files may be submitted by email, FTP, or on diskette. The DVI file(s) and PostScript files should be submitted only by FTP or on diskette unless they are encoded properly to submit through email. (DVI files are binary and PostScript files tend to be very large.)

Electronically prepared manuscripts can be sent via email to pub-submit@ams.org (Internet). The subject line of the message should include the publication code to identify it as a Memoir. TeX source files, DVI files, and PostScript files can be transferred over the Internet by FTP to the Internet node e-math.ams.org (130.44.1.100).

Electronic graphics. Comprehensive instructions on preparing graphics are available at www.ams.org/jourhtml/graphics.html. A few of the major requirements are given here.

Submit files for graphics as EPS (Encapsulated PostScript) files. This includes graphics originated via a graphics application as well as scanned photographs or other computer-generated images. If this is not possible, TIFF files are acceptable as long as they can be opened in Adobe Photoshop or Illustrator. No matter what method was used to produce the graphic, it is necessary to provide a paper copy to the AMS.

Authors using graphics packages for the creation of electronic art should also avoid the use of any lines thinner than 0.5 points in width. Many graphics packages allow the user to specify a "hairline" for a very thin line. Hairlines often look acceptable when proofed on a typical laser printer. However, when produced on a high-resolution laser imagesetter, hairlines become nearly invisible and will be lost entirely in the final printing process.

Screens should be set to values between 15% and 85%. Screens which fall outside of this range are too light or too dark to print correctly. Variations of screens within a graphic should be no less than 10%.

Inquiries. Any inquiries concerning a paper that has been accepted for publication should be sent directly to the Electronic Prepress Department, American Mathematical Society, P. O. Box 6248, Providence, RI 02940-6248.

Editors

Selected Titles in This Series

(Continued from the front of this publication)

694 **Alberto Bressan, Graziano Crasta, and Benedetto Piccoli,** Well-posedness of the Cauchy problem for $n \times n$ systems of conservation laws, 2000

693 **Doug Pickrell,** Invariant measures for unitary groups associated to Kac-Moody Lie algebras, 2000

692 **Mara D. Neusel,** Inverse invariant theory and Steenrod operations, 2000

691 **Bruce Hughes and Stratos Prassidis,** Control and relaxation over the circle, 2000

690 **Robert Rumely, Chi Fong Lau, and Robert Varley,** Existence of the sectional capacity, 2000

689 **M. A. Dickmann and F. Miraglia,** Special groups: Boolean-theoretic methods in the theory of quadratic forms, 2000

688 **Piotr Hajłasz and Pekka Koskela,** Sobolev met Poincaré, 2000

687 **Guy David and Stephen Semmes,** Uniform rectifiability and quasiminimizing sets of arbitrary codimension, 2000

686 **L. Gaunce Lewis, Jr.,** Splitting theorems for certain equivariant spectra, 2000

685 **Jean-Luc Joly, Guy Metivier, and Jeffrey Rauch,** Caustics for dissipative semilinear oscillations, 2000

684 **Harvey I. Blau, Bangteng Xu, Z. Arad, E. Fisman, V. Miloslavsky, and M. Muzychuk,** Homogeneous integral table algebras of degree three: A trilogy, 2000

683 **Serge Bouc,** Non-additive exact functors and tensor induction for Mackey functors, 2000

682 **Martin Majewski,** ational homotopical models and uniqueness, 2000

681 **David P. Blecher, Paul S. Muhly, and Vern I. Paulsen,** Categories of operator modules (Morita equivalence and projective modules, 2000

680 **Joachim Zacharias,** Continuous tensor products and Arveson's spectral C^*-algebras, 2000

679 **Y. A. Abramovich and A. K. Kitover,** Inverses of disjointness preserving operators, 2000

678 **Wilhelm Stannat,** The theory of generalized Dirichlet forms and its applications in analysis and stochastics, 1999

677 **Volodymyr V. Lyubashenko,** Squared Hopf algebras, 1999

676 **S. Strelitz,** Asymptotics for solutions of linear differential equations having turning points with applications, 1999

675 **Michael B. Marcus and Jay Rosen,** Renormalized self-intersection local times and Wick power chaos processes, 1999

674 **R. Lawther and D. M. Testerman,** A_1 subgroups of exceptional algebraic groups, 1999

673 **John Lott,** Diffeomorphisms and noncommutative analytic torsion, 1999

672 **Yael Karshon,** Periodic Hamiltonian flows on four dimensional manifolds, 1999

671 **Andrzej Rosłanowski and Saharon Shelah,** Norms on possibilities I: Forcing with trees and creatures, 1999

670 **Steve Jackson,** A computation of δ^1_5, 1999

669 **Seán Keel and James McKernan,** Rational curves on quasi-projective surfaces, 1999

668 **E. N. Dancer and P. Poláčik,** Realization of vector fields and dynamics of spatially homogeneous parabolic equations, 1999

667 **Ethan Akin,** Simplicial dynamical systems, 1999

666 **Mark Hovey and Neil P. Strickland,** Morava K-theories and localisation, 1999

665 **George Lawrence Ashline,** The defect relation of meromorphic maps on parabolic manifolds, 1999